William Baird, Linnean Society of London

Description of a New British Annelide

belonging to the tribe Rapacea of Grube Annelida errantia of

Milne-Edwards

William Baird, Linnean Society of London

Description of a New British Annelide
belonging to the tribe Rapacea of Grube Annelida errantia of Milne-Edwards

ISBN/EAN: 9783337225162

Printed in Europe, USA, Canada, Australia, Japan

Cover: Foto ©Andreas Hilbeck / pixelio.de

More available books at **www.hansebooks.com**

[*Extracted from the* LINNEAN SOCIETY'S JOURNAL.—ZOOLOGY, vol. viii.

DESCRIPTION

OF A

NEW BRITISH ANNELIDE,

BELONGING TO THE

TRIBE RAPACEA of GRUBE
= *Annelida errantia* of MILNE-EDWARDS.

BY

W. BAIRD, M.D., F.L.S.

about one-third the whole length and is composed of 21 segments, the posterior containing 85. In form it tapers gradually from the head to the tail, which terminates in a blunt point without cirri. The dorsal surface is beautifully marked with dark-purple spots, which extend over the upper part of the feet, leaving a hollow space in the centre free from them. The anterior portion of the body is convex, the lower flattened. The segments of the anterior part are of considerable size, but those of the lower are small and very much crowded together. A canal runs down the ventral surface the whole length, while a similar one runs down the dorsal surface of the anterior portion only, beginning at the sixth or seventh segment and continuing to the twenty-first. The head is rather small; the first segment of the body of moderate size, and the four succeeding ones very narrow (fig. 1 d), the first nearly equalling the three following. The tentacular cirri are equal in length to about the four first segments. The feet of the anterior portion of the body (fig. 1 a) are all simple lobulated feet, without any foliaceous branchial lamina. The dorsal lobe is short, stout, and rounded at the apex, with a cirrus springing from near its root, about one-third longer than the lobe itself, and not crenated underneath. The ventral lobe is somewhat larger than the dorsal, and the cirrus attached to it is very short, not quite half the length of the lobe. The bristles attached to it are of two kinds: those especially attached to the lobe nearest the dorsal lobe (the superior fascicle) are all slender, compound, with a sharp-pointed smooth style inserted into a stouter cylindrical shaft which is slightly striated (*setæ spinosæ*). The bristles of the inferior branch are bifasciculate, and consist of one bundle composed of spinous bristles like those above described, and another composed of stouter and rather shorter setæ with a striated slightly bent cylindrical shaft cut obliquely at the tip, to which portion is articulated a short claw-like piece, quite smooth, and slightly bent at the apex (*setæ falcatæ*). The aciculæ are stout and of a black colour. The posterior feet (fig. 1 b) are all much smaller than the anterior, but more complicated in structure. Above the base of the dorsal or superior lobe we find a compressed foliaceous lamella. The superior setiferous tubercle is also furnished with a similar lamina, but very large, extending across the middle lobes of the foot. The ventral cirrus has appended beneath its base another foliaceous lamina of about the same size as that attached to the dorsal lobe. The cirri of these lobes are rather short, that of the dorsal lobe being longer than the lobe, while the ventral cirrus is shorter than its corresponding lobe. The bristles are of two kinds,—those of the setiferous tubercles being numerous, compound, and consisting of a flattened lancet-shaped blade, smooth and rather sharp-pointed, let into a somewhat cylindrical shaft which is striated half across (*setæ cultratæ*) (fig. 1 *c*). Mixed with these are a few (about four or five) long and stout setæ of the falcate kind, but much longer and stronger than those of the anterior feet.

The species which approaches nearest to this is the *Heteronereis renalis* of Johnston = *H. arctica* of Oersted. It differs, however, in many respects. The relative size of the first and four succeeding segments, the colour and peculiar markings of the body, the canal running down the centre dorsally and ventrally, the number of anterior segments (in *renalis* or *arctica* being only twenty, while in this species there are twenty-one), the posterior portion of the body being more slender, and the tail destitute of cirri, the structure of the feet and cirri, &c.,—all separate it from that species.

The only specimen which I have seen was found by Mr. Laughrin at Polperro, Cornwall, in a muddy bottom, and is now in the national collection, British Museum.

Description of several new Species and Varieties of Tubicolous Annelides = Tribe LIMIVORA of Grube, in the Collection of the British Museum. By W. BAIRD, M.D., F.L.S.—PART I.

(Plate I.)

[Read April 21, 1864.]

IN the extensive collection of Annelides belonging to the British Museum, now in course of arrangement, there is a considerable number which appear to me to be undescribed. In many cases these are difficult to determine, from the fact that soft animals preserved in spirits do not always retain their form and consistence, or may be so hardened by the spirits as to offer great difficulties in making out the different parts. In the case of the Tubicolous Annelides, again, perhaps only the tubes or cases in which the animals dwell have been preserved, and thus it is almost impossible to refer them to their proper genera. As, however, notwithstanding these difficulties, we possess many species which can be determined, it is my intention, from time to time, to offer to the Linnean Society, if approved of, descriptions of such species as appear to me new or worthy of particular attention.

Family SERPULIDÆ.

The genus *Serpula* of Linnæus, as established by the illustrious Swede, contained several species now known to belong to the genus *Vermetus*, a genus of mollusks. After these were withdrawn, there still remained many forms of shelly tubes which, though bearing a general resemblance to each other, were difficult to be arranged under one single genus. The animals, however, the architects of these tubes, after a time began to be a little more studied; and thus Lamarck, Blainville, Savigny, and some other naturalists were enabled to construct, upon good grounds, several genera to contain what might otherwise have been considered similar forms. The last author who has paid particular attention to this Linnean genus is Dr. A. Philippi. His subdivisions of *Serpula* are founded upon a character which has been discovered by malacologists to be of great value in the class Mollusca. The animals of the greater number of the species of *Serpula* which have been described possess a similar organ to that which characterizes so many of the Gasteropodous Mollusca. This is the operculum, which varies considerably in structure in the different species, and which thus forms an excellent character for dividing them into genera. As Philippi justly observes, " this character has, moreover, the advantage that it may still be frequently observed in dried specimens preserved in museums."

Little dependence can be placed on the *shelly tube* alone in distinguishing the species or even the genera: thus we find a similar shell possessed by two or three different Annelides belonging to two or three distinct genera; for, as Philippi remarks in his paper*, " the shells of *Serpula triquetra*, *Vermilia triquetra*, and *Pomatoceros tricuspis* are difficult to distinguish without the animals."

The structure of the operculum is far more varied, indeed, than it had been hitherto supposed to be; and I think Dr. Philippi has done good service to the students of this group of Annelides by so carefully distinguishing the structure of this appendage. It is owing to the fact mentioned above (that the operculum frequently

* Wiegmann's Archiv for 1844, Band i. p. 186. Translated into English by Dr. Francis, in Ann. & Mag. of Nat. Hist. 1844, vol. xiv. p. 153-162.

remains behind in dried specimens), that I have been enabled to add some new species, belonging to the national collection, not hitherto described. The number of genera characterized by Philippi belonging to the Serpulidæ is ten, and the species enumerated by him as occurring in the Mediterranean alone are twenty-five. Various other exotic species have been described at different times, and to these I now propose adding several more.

Genus EUPOMATUS, *Philippi* *.

1. EUPOMATUS BOLTONI, Baird. (Pl. I. figs. 2, 2 *a*, *b*.)

Char. Animal (operculo excepto) ignotum. Operculum corneum, infundibuliforme, margine externo dense crenato, interne cuspidibus calcareis viginti dentatis instructum. Testa rubra, triquetra, adhærens, transversim rugosa, dorso canaliculata.

Hab. Nova Zelandia. (Mus. Brit.)

This is a fine species of the family Serpulidæ, of which, however, we have as yet only received the shelly tube and the operculum of the animal. In our national collection we possess three good specimens of the shell and three specimens of the operculum. This portion of the animal is large, and by means of it we can distinctly refer the species to the genus *Eupomatus* of Philippi. It is rounded, slightly funnel-shaped, and of a horny texture (Pl. I. fig. 2 *a*). Externally the margin is densely crenated—the crenations being about eighty-eight or ninety in number, and tooth-like. Internally it is provided with a considerable number (about twenty) of hard, flattened, calcareous spikes (or, as Philippi elsewhere calls them, horns, *cornua*), rising up from the centre and strongly dentate—these teeth being four or five in number, stout, rather blunt, and arranged on one side only (fig. 2 *b*). The spike itself terminates in a claw-shaped sharp point, slightly curved at the extremity. These spikes bear altogether an exact resemblance to the toothed extremity of the large claw of a lobster. The tube, in all the specimens which I have seen, is found attached to, and creeping on, dead shells (fig. 2). In one specimen, which, however, is not quite perfect at the posterior extremity, it is about three inches in length. It is of a red colour, triquetrous where attached, but round at the anterior ex-

* The genus *Eupomatus* was constituted by Philippi to receive those species of *Serpula* that had the operculum furnished on the upper side, in the centre, with a certain number of moveable spikes. The operculum, he says, is horny, and in the Mediterranean species these spikes are horny also; but this latter character does not hold good in all the other species which have been described.

tremity or mouth when the tube raises itself up from the shell upon which it creeps, is corrugated transversely (the striæ of growth?), and is marked with a large, distinct canal or furrow, running along the dorsal surface throughout its whole length.

Of the three specimens we possess, one, the largest, is attached to part of the shell of *Haliotis australis*, another to a fragment of a species of *Mactra*, and the third is coiled round a species of *Elenchus*.

They were all collected in New Zealand by Lieut.-Col. Bolton, R.E., to whom I have dedicated the species.

Genus PLACOSTEGUS *, *Philippi*.

2. PLACOSTEGUS CARINIFERUS, *Gray* (sp.), *Baird*.

Numerous specimens of this species of Annelide were brought at different times from New Zealand, and deposited in the national collection, by the late lamented Dr. Andrew Sinclair, R.N., Lieut.-Col. Bolton, R.E., the late Captain Sir Everard Home, Bart., and His Excellency Governor Sir George Grey.

The tube or shell was briefly described by Dr. Gray in 1843, in the 'Fauna of New Zealand' appended to Dr. Dieffenbach's 'Travels in New Zealand.' As only the operculum was known at that time to Dr. Gray, and as that resembles very much in form the operculum of the molluscous genus of shells "*Vermetus*," he described it under the name of *Vermetus cariniferus*. A similar, and, I believe, the identical species has since that time been described and the animal figured by Schmarda, in his 'Neue wirbellose Thiere,' 1861, under the name of *Placostegus cœruleus*. My chief object in this brief notice is to give a few more particulars with regard to this species, to correct the synonymy, and to restore the specific name attached to it originally by Dr. Gray. I wish also particularly to bring before the notice of the Society the fact that the animal gives out a beautiful dye or colour. The specimens which were the subjects of my examination had been for a number of years in the British Museum, some having been placed there in 1845, and others in 1847. Notwithstanding their having been so long dry, when softened in water, taken out of the tubes, and placed in spirits of wine, they imparted to the

---

* The genus *Placostegus* was constituted by Philippi to contain those species of *Serpula* which have a calcareous operculum (approaching very nearly in form to that of some of the Gasteropodous Mollusca) in the shape of a shallow disk, entire at the margin.

liquid a beautiful and delicate red tint. The whole animal is of a fine blue colour, and the elegant tuft of branchial filaments intensely azure banded with white. In describing the tube of this species of Annelide in 1843, Dr. Gray had only one or two specimens to describe from, as the other specimens, which are now in the Collection of the British Museum, arrived long after that description was drawn up. He says, "the shell is thick, irregularly twisted, opaque white, with a high compressed wavy keel along the upper edge; mouth orbicular, with a tooth above it, formed by the keel. Operculum orbicular, horny." In the collection there are two or three specimens which occur single, and were found creeping on dead shells. To these this description applies very well; but, in addition to those, we have various specimens collected together into large masses nearly the size of a small human head, and consisting of several thousands of tubes twisted and twined together. In the generality of these we see the keel, mentioned by Dr. Gray as "high," "compressed," and forming " a tooth " at its extremity, becoming double as it were at a certain distance from the mouth of the tube, diverging a little from each other, the surface of the tube between the two keels being raised to the same height as the tube, and thus forming a rather broad flat tooth or strap which projects considerably beyond the circular rim of the mouth. In many specimens this tooth is sharp-pointed, but in others it is blunt and rounded at the point.

Schmarda asserts that the species described by him is also a native of the Cape of Good Hope. His description applies better to the New Zealand specimens than to those from the Cape, and I was led at first to separate the two as distinct species. A more careful examination, however, of all the specimens we possess from both these habitats, has now induced me to consider those from the Cape of Good Hope to be only a variety of the other. Several specimens of this variety, occurring in large masses of some thousands of tubes clustered together, were collected by Dr. Krauss many years ago at the Cape of Good Hope, and are now in the Collection of the British Museum.

This variety I have named

PLACOSTEGUS CARINIFERUS, var. *Kraussii*;
and I here append a more detailed description of it.

 Char. Animal *Placostego carinifero* valde simile, sed minus intense cæruleum. Branchiæ pallide cæruleæ, albo-fasciatæ, filamentis circiter viginti et sex, uno latere plumosis. Setæ pedum longæ, numerosæ, simplices, ad finem curvatæ. Tubuli repentes, in massam magnam glomerati, dorso plane carinati, ligula plana, os supra extensa terminati.

Hab. Promontorio Bonæ Spei. (Mus. Brit.)

The animal differs from that of the specimens from New Zealand in being less deeply coloured, and perhaps being longer in proportion to the size of the tube. This is smaller, and the dorsal keel is perhaps rather flatter and less sharp-pointed at its extremity. The two sets of specimens, however, agree in this particular, that the animals, when softened in water and then immersed in spirits of wine, impart to the liquid the same beautiful red colour, though, as may be supposed from the animal being less deeply coloured, those from the Cape of Good Hope give out a slightly fainter hue.

3. PLACOSTEGUS LATILIGULATUS, *Baird.* (Pl. I. figs. 3, 3*a*, *b*.)

Char. Animal *Placostego carinifero* simile. Color corporis fuscus. Branchiæ albæ, cæruleo fasciatæ. Operculum calcareum, circulare, concavum, cæruleum. Tubuli repentes, flexuosi, dorso late carinati, carina in latam ligulam, supra os extensam desinens. Os interne cæruleum.

Hab. ——? (Mus. Brit.)

Only one mass, consisting of about 100 or more tubes, is in the possession of the Museum, and no history is attached to the specimen. The animal, softened in water and taken out of the tube, as far as can be ascertained from the imperfect state of the specimens, is very similar in appearance to the animal of the *Placostegus cariniferus.* It is about the same size as those taken from the var. *Kraussii,* from the Cape of Good Hope, but differs a good deal in colour. The body of the animal is of a fuscous-brown colour, the branchial filaments white, banded with blue, and the operculum is of an azure hue. The tubes are broad, clustered together, and creeping in a very flexuous manner; they are of a bluish colour, the mouth of the tube deeply so, and the flat dorsal keel is somewhat of the same hue. The tube itself and the keel which runs along the back are broad, the latter part especially so at its extremity, where it terminates in a flat, strap-like tooth or sort of hood which extends some way beyond the rounded mouth (fig. 3 *b*). The surface throughout is much wrinkled, and the whole tube presents an irregular form of growth.

We have no history attached to this specimen; and were it not that the animals in some of the tubes still exist, the mass might be taken for a group of fossil tubes.

4. PLACOSTEGUS GRAYI, *Baird.* (Pl. 1. figs. 4, 4 *a*, *b*.)

Char. Animal, operculo excepto, ignotum. Operculum corneum?, circulare, concavum. Tubuli flexuose repentes, depressi, valde rugosi, dorso late carinati, carina haud in ligulam os supra extensam desinens.

Hab. ——? (Mus. Brit.)

The only specimens we possess in the collection of the British Museum are a few tubes creeping on a stone. The operculum was found in two or three of the tubes, and, unlike the others belonging to the genus *Placostegus*, appears to be horny, of a circular form, and hollow or concave on its upper surface. The tubes are flexuose, very rugose, and possess, like the last-described species (*P. latiligulatus*), a rather broad flat keel along the back of the shell. This keel is very rugose or wrinkled, and does not extend beyond the mouth of the tube, which is quite circular (fig. 4 *b*). The form of the tube is very irregular, and in several specimens at the larger extremity it is cemented as it were by a smooth, hard calcareous secretion to the stone to which it is attached. The specimens were presented many years ago to the Museum by Dr. Gray, whose name I have attached to the species.

EXPLANATION OF PLATE I.

Fig. 1. *Heteronereis signata*, natural size; 1 *a*, one of anterior feet; 1 *b*, one of posterior feet; 1 *c*, seta of ditto; 1 *d*, head and 8 first segments of body: all magnified.

Fig. 2. *Eupomatus Boltoni*, natural size, on *Haliotis*; 2 *a*, operculum of ditto; 2 *b*, one of the spikes of ditto: both magnified.

Fig. 3. *Placostegus latiligulatus*, nat. size; 3 *a*, operculum of ditto: magnified; 3 *b*, extremity of tube, nat. size.

Fig. 4. *Placostegus Grayi*, nat. size; 4 *a*, operculum of ditto: magnified; 4 *b*, extremity of tube, nat. size.

PART II.

(Plate II.)

[Read June 2, 1864.]

Genus CYMOSPIRA, *Savigny.*

Amongst the tubicolous Annelides belonging to the family Serpulidæ, the genus *Cymospira* of Savigny is remarkable. The branchiæ are described by Pallas and others as being very beautiful when seen in the living animal, and are rolled into spires of several turns. The operculum consists of a somewhat horny, elliptical, shallow plate, which supports two or more dentated horns or processes, generally near its hinder margin. The tubes of all the known species, of which only three or four have been described, burrow into or are attached to masses of Madrepore in the seas of the West Indies. In the collection of Annelides belonging to the British Museum we possess several additional species, found inhabiting coral in other parts of the world. One of these was found on a coral reef in the Arabian Gulf, and, in the structure of the operculum, &c., materially differs from all that have been previously described. The following is its description:—

5. CYMOSPIRA TRICORNIS, *Baird.* (Pl. II. fig. 1, operculum.)

Branchiæ in spiras quinque convolutæ. Operculum magnum, cornibus tribus dentatis armatum.

The branchiæ are disposed in five whorls. The filaments are densely plumose on one side and are of moderate length. The operculigerous filament is thick and fleshy. The operculum is large, nearly flat on the upper surface, and is armed with three stout, irregularly-toothed horns. The collar is large and fleshy. The spines of the thoracic segments are stout, rather short, and yellowish-coloured. The abdominal portion of the body is about 2 inches long, smooth on the ventral surface with the exception of a few longitudinal strong striæ, and strongly and densely striated across on the dorsal surface. The tube in which this annelide dwells is large, nearly as thick as a man's little finger, but so covered with coral deposit that it is very difficult to ascertain its form. We possess in the British Museum only two specimens of this animal, one of them being partly contained in a fragment of its tube. The mouth of this tube seems to be nearly round; but the rest of it is so covered with madrepore, in a mass of which it had apparently burrowed, that nothing more can be seen of its structure.

The whole animal is fully 3 inches long, tapered somewhat towards the tail, and about the centre of the body is nearly 4 lines in diameter.

Hab. Djedda, in coral reef. From the Collection of Mr. Metcalf. (Brit. Mus.)

6. CYMOSPIRA BRACHYCERA, *Baird.* (Pl. II. fig. 2, operculum.)

Branchiæ in spiras quinque convolutæ. Operculum magnum, cornibus duobus brevissimis irregulariter dentatis armatum.

Amongst the numerous objects of natural history collected during the surveying-voyage of H. M. S. 'Fly' by Mr. Jukes, Naturalist to the Expedition, and transmitted by him to the British Museum, are two specimens from Swain's Reefs, on the east coast of Australia, of the " animals of tubes that bore into coral." Neither the tubes themselves, nor fragments of the coral containing them, were secured; but as no doubt the former, like the other known species, would be completely immersed in and incrusted by the latter, little information could be obtained from them.

The branchiæ are coiled round in five spires. The filaments are of moderate length, and plumose on one side. The collar is rather thin and membranous. The operculigerous filament is thick and fleshy, and the operculum itself is large, of an oval form, and armed on its slightly concave surface with two very short and irregularly-toothed horns. The thoracic portion of the body is short and rather square-shaped, with a free margin on each side and on the lower edge; and the setæ of the feet are rather short and bright yellow. The abdominal portion is strongly and densely striated across. The entire length of the animal is about 3 inches (in spirits).

Hab. East coast of Australia. (Brit. Mus.)

The way in which these animals were seen and collected is thus described by Mr. Jukes in his Narrative of the voyage:—
"A block of coral rock that was brought up by a fish-hook from the bottom at one of our anchorages was interesting from the vast variety and abundance of animal life there was about it. It was a mere worn, dead fragment; but its surface was covered with brown, crimson, and yellow nulliporæ, many small actiniæ and soft branching corallines, sheets of flustra and eschara, and delicate reteporæ, looking like beautiful lacework carved in ivory. There were several small sponges and alcyonia, sea-weeds of two or three species, two species of comatula and one of ophiura of the most delicate colours and markings, and many small, flat, round corals, something like nummulites in external appearance. On breaking into the block, boring shells of several species were found buried in it; *tubes formed by Annelida* pierced it in all directions, many still containing their inhabitants, while two or three worms, or nereis, lay twisted in and out among its hollows and recesses, in which, likewise, were three small species of crabs. This block was not above a foot in diameter, and was a perfect museum in itself, while its outside glared with beauty from the many brightly and variously coloured animals and plants. It was by no means a solitary instance; every block that could be procured from the bottom, in from 10 to 20 fathoms, was like it. What an inconceivable amount of animal life must be here scattered over the bottom of the sea, to say nothing of that moving

through its waters, and this through spaces of hundreds of miles! Every corner and crevice, every point occupied by living beings, which, as they become more minute, increase in tenfold abundance." (p. 17.)

In the same collection of Annelides we possess specimens of a tube imbedded in madrepore collected by Mr. John MacGillivray from the coral reef of the island of Totoga, one of the Fiji group. From its appearance and habitat I consider it to belong to the same genus as the last, and propose naming it.

7. CYMOSPIRA MACGILLIVRAYI. (Pl. II. fig. 3, mouth of tube.)

Only the mouth of the tube is distinctly seen, the remainder being imbedded in and completely incrusted by the substance of the madrepore. The mouth of the tube is round, smooth internally but of a dark colour tinged with red, and at the upper edge is strongly marked with the projecting point of a keel, which most probably runs along the dorsal surface of the tube. This projecting point is somewhat tongue-shaped, of a smooth surface and a reddish colour, and reflected a little upwards and backwards.

It is to be regretted that the specimens we possess are so few in number, and the fragments of the madrepore which contain the tubes so small that it is impossible to ascertain the length of the tube. The circumference of the mouth of the largest specimen is fully ⅜ths of an inch.

Hab. Coral reef of Totoga, Fiji Islands. (Brit. Mus.)

Genus POMATOSTEGUS, *Schmarda.*

When Philippi reconstructed the family Serpulidæ, taking the structure of the operculum as one of his chief generic characters, only two species of the genus *Cymospira* had then been described. One of these, the type of the genus, was the *Serpula gigantea* of Pallas,=the *Terebella bicornis* of Abildgaard, distinguished by its having an operculum consisting of an elliptical shallow plate armed with two ramified horns. The other was the *Terebella stellata* of Abildgaard, distinguished by the operculum being as it were multiplied, or raised up in three different floors or stories united to each other by a central column. Following up the subdivisions of Philippi founded on the operculum as a character, Schmarda has since founded a new genus for this latter annelide, which he has called *Pomatostegus*, and has described two new species from the coral reefs of Jamaica. The worm which I have now to describe belongs to this genus, but is a native of the seas of Australia. A single specimen was added to our collection about eight or nine months ago by Dr. Bowerbank, but no tube was collected, nor have we any further information about it.

8. POMATOSTEGUS BOWERBANKI, *Baird*. (Pl. 11. figs. 4 & 5, operculum.)

Branchiæ curtæ, in spiram unam et dimidiam convolutæ. Opercula quatuor, versus apicem decrescentia, inarmata.

The branchiæ are rather short, the filaments plumose on one side only. Operculigerous lobe thick and fleshy. Operculum consisting of four stories united by a common central column, and densely covered with a rough coat of short hairs or filaments of a fibrous substance. These opercula diminish in size as they ascend, the last being very small and not armed with any spines or horns. Collar small. Thoracic portion of body short, square-shaped. Bristles of feet rather long and of a yellowish colour. Abdominal portion of body gradually tapering to the extremity, and striated across, the striæ wide apart. It is of a reddish-brown colour. The total length is $2\frac{1}{2}$ inches.

Hab. Seas of Australia. (Brit. Mus.)

Genus SERPULA, *as restricted by Philippi.*

Taking the operculum as his principal character, Philippi restricts the old genus *Serpula* to those species which are distinguished by having the operculum of a horny substance, in the form of a rather shallow or funnel-shaped plate, the concave disk crenate on the margin, radiately grooved above, and supported on a subconical fleshy petiole. This organ is in many species of a beautiful shape, and, having in some instances a vitreous look, might, as Dr. Johnston well observes, "make an elegant pattern for a wineglass."

The species hitherto enumerated have been confined to the European fauna: I am not aware, at least, of any that have been described from any other part of the world; and Schmarda, who is amongst the latest authors that have paid attention to exotic Annelides, remarks that, however common they are in the Mediterranean, he has not found one exotic species. It is with much pleasure, therefore, that I dedicate the following, from Australia, to Mr. Jukes, to whom the British Museum is indebted for the specimen.

9. SERPULA JUKESII, *Baird*. (Pl. 11. fig. 6, operculum.)

Branchiæ in spiram unam convolutæ, lacteæ, filamentis dorso canaliculatis. Operculum et filamentum operculigerum alba. Operculum profunde infundibulatum, multicrenatum. Tubus teres, solidus.

A single specimen of this species of the restricted genus *Serpula* was taken by Mr. Jukes, during the voyage of the 'Fly,' on the coast of Australia. The branchiæ are rolled up in a single spire, and are of a dull milk-white colour; the filaments are about thirty-four in number on each side, and on the dorsal surface are pretty deeply grooved or channelled. The operculum is deeply infundibuliform, of a white colour, the edge indented with numerous close-set crenations, the grooves extending down along the whole length of the outer surface. The body of the animal tapers towards the extremity, and is of a dull reddish colour and strongly striated across. The total length of the animal is about $1\frac{1}{2}$ inch, the breadth about $1\frac{1}{2}$ line. Only a fragment of the tube in which the worm lives was preserved. It is perfectly cylindrical, without any keel or striæ, is thick and solid, and of a white colour externally.

Hab. Seas of Australia. (Brit. Mus.)

10. SERPULA NARCONENSIS, *Baird*. (Pl. II. figs. 7 & 8, operculum.)

> Branchiæ in spiram unam convolutæ. Operculum lacteum, minime profundum, dense crenatum; petiolum operculigerum gracile, prope finem nodosum.

This is a small species collected at Narcon Island during Captain Sir J. Ross's Antarctic exploring expedition; and only one specimen, without the tube, was procured. The chief character which marks the species is the form of the operculum. This is a white, rather shallow disk, elegantly formed, beautifully multicrenate on the margin, and radiately grooved on its upper surface internally as well as externally. The pedicle which supports it is slender, and terminates near the summit in a rounded knob, upon which the operculum is seated, being attached to it by a short stalk, which appears like a moveable joint. There is nothing particular in the form or characters of the body, except that it is short and stout, measuring in total length, including branchiæ and operculum, about 10 lines.

Hab. Narcon Island. (Brit. Mus.)

11. SERPULA ZELANDICA, *Baird*. (Pl. II. fig. 9, operculum.)

> Animal, operculo excepto, ignotum. Operculum album, parvum, minime profundum, margine crenis viginti ornatum. Tubus gracilis, albus, repens, fere rotundus, carina longitudinali parva in dorso signatus; transversim flexuose striatus.

Several specimens of this small species of *Serpula* are in the collection of the Museum, the slender tubes creeping on fragments of old oyster-shells. The operculum is the only part of the animal preserved, as the specimens were transmitted in a dry state. Like that of the other known species of true *Serpula*, it is finely crenated on the margin. The crenæ are twenty in number, but the grooves externally are confined to the surface of the disk itself, and are not extended to the pedicel or stalk. The tube is slender, nearly round, with only a slight keel running longitudinally along its dorsal surface. It is white, the mouth is nearly circular, and the shell itself is strongly marked along its whole length with transverse flexuous striæ which encircle it.

The specimens in the collection are grouped together on the old oyster-shell, and mixed up with numerous specimens of zoophytes, Alcyonia &c. Most of them are more or less incrusted with these substances. Length of the tube about 16 lines; circumference about 1 line.

Hab. New Zealand. (Brit. Mus.)

EXPLANATION OF PLATE II.

Fig. 1. *Cymospira tricornis*, operculum.
2. *C. brachycera*, operculum.
3. *C. MacGillivrayi*, mouth of tube, in coral.
4, 5. *Pomatostegus Bowerbanki*, operculum.
6. *Serpula Jukesii*, operculum.
7, 8. *S. Narconensis*, operculum.
9. *S. Zelandica*, operculum.
10. *Eupomatus Boltoni*, operculum.
11. *Galeolaria decumbens*, operculum.

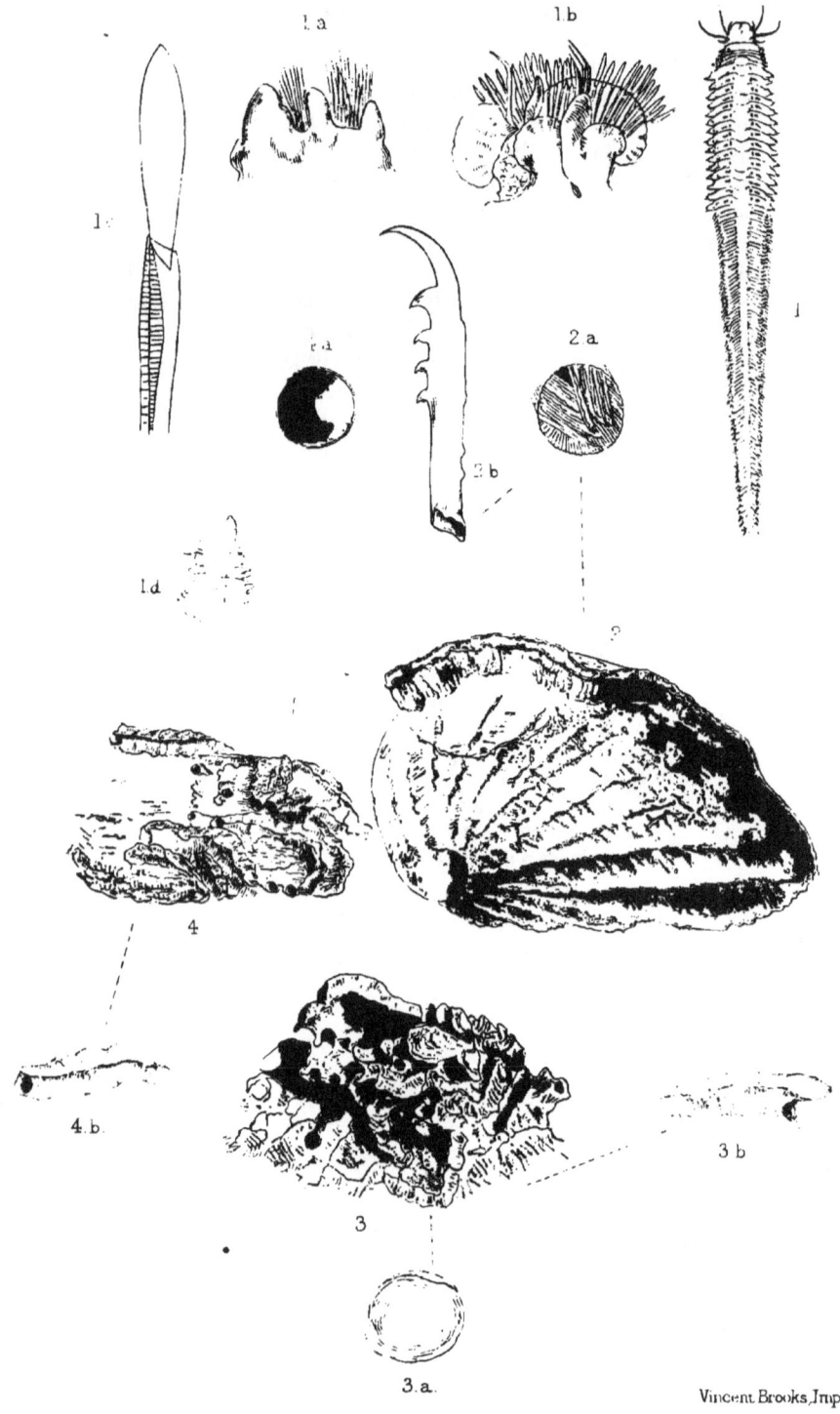

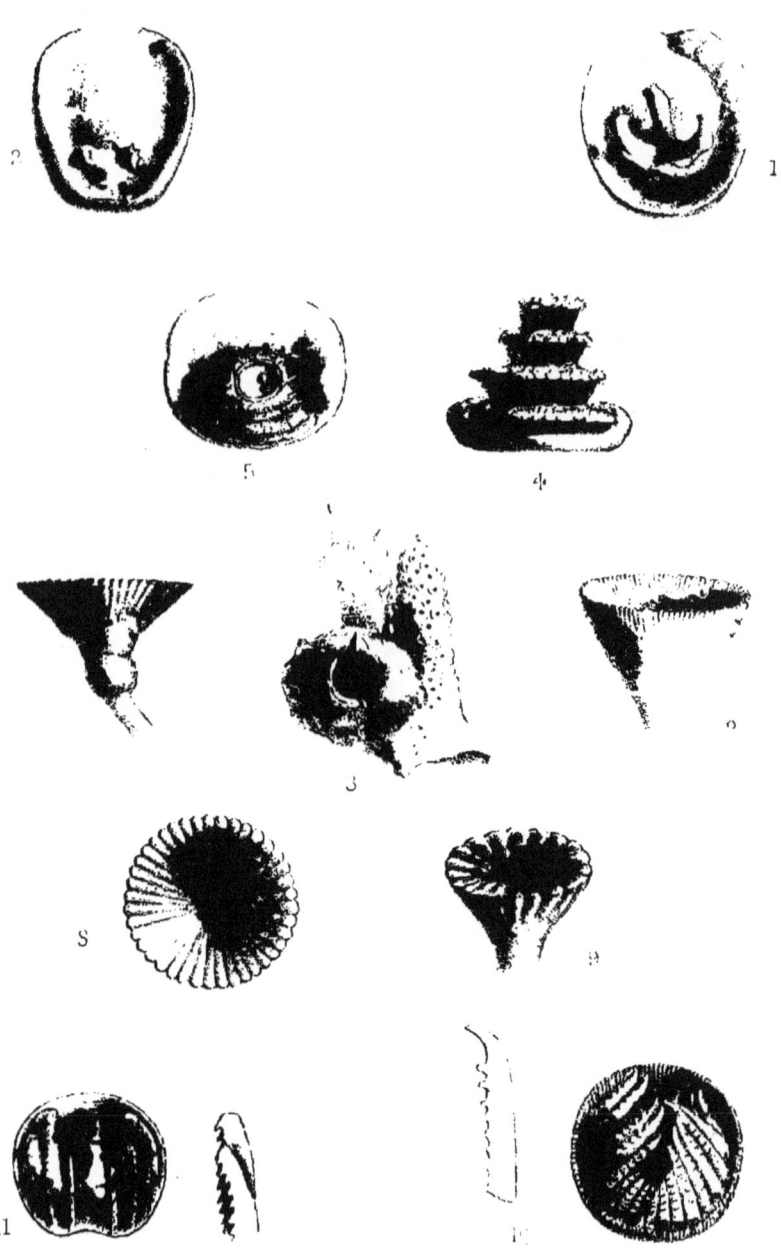

On new Tubicolous Annelids, in the Collection of the British Museum. Part 2. By W. Baird, M.D., F.L.S.

[Read December 1, 1864.]

[Plate V.]

Genus Terebella, *Linn.* (*Montagu*).

1. Terebella flabellum, *Baird.* Pl. V. figs. 1 & 2.

The animal is as yet unknown, but the tubes are sufficiently remarkable to merit a description. The specimens, which have been deposited in the collection of the British Museum, vary in size, the largest being about 6 inches in length, and about the circumference of an ordinary goose-quill. They are cylindrical in form, tapering gradually from the summit to the base; the upper portion being the narrower. They are composed of a thin membranous substance internally, covered externally with numerous fragments of shells, corals, and pieces of horny zoophytes. The most characteristic feature, however, in the structure of this tube is the fan-shaped expansion of filaments at its upper orifice. This orifice is circular, and has on its dorsal surface a projecting lip or kind of hood (fig. 2), which extends beyond the mouth for a short distance, whilst from its ventral side springs another lip or hood (fig. 1), which quickly expands into a fan-shaped tuft of horny-looking filaments. This tuft is composed of several branches, each of which divides dichotomously into stiff but somewhat flexible filaments, spreading out horizontally to the length of an inch or more. These filaments are nodulous, and seem to possess a glutinous secretion, by means of which they are able to attach small shells, &c. to their surface.

Hab. These tubes, to the number of six, were collected during Sir J. Clarke Ross's Antarctic Expedition—two of them being registered in our collection as from *Narcon* Island. (Mus. Brit.)

2. Terebella bilineata, *Baird.* Pl. V. figs. 3 & 4.

Animal with three pairs of branchiæ, composed of simple cirriform filaments (fig. 3). They are not arborescent, having no trunk or main branch from which the others spring, but are inserted in tufts of single filaments on the three first segments of the body, on each side. The tentacula are composed of numerous, rather long filaments, hollowed in the centre, and waved or undulated along the edge. In the specimens we have preserved in the collection, most of these have unfortunately fallen off. The bristle tufts are continued to the end of the body, and are about 36 in number. The segments of the body are rather deeply striated across, and the surface is somewhat granular in appearance. The body is thickest about the centre, and tapers suddenly from that to the inferior extremity. When alive, the animal is marked with two fine stripes or lines running longitudinally down along the dorsal surface beautifully tinged with purple.

The case or tube which this animal constructs (fig. 4), and which it inhabits, is of an irregular form, and consists of a thin transparent membrane, densely coated externally with numerous rough fragments of stones and shells, with some beautiful foraminifera mixed, coarsely cemented together and exhibiting a very rude appearance.

The total length of the animal is about 3 inches, and that of the longest tube is about 4 or $4\frac{1}{2}$ inches.

Hab. Falkland Islands. Collected by Mr. W. Wright. (Mus. Brit.)

Genus SABELLA, *Linn.* (*Savigny*).

1. SABELLA BIPUNCTATA, *Baird.*

Worm rather slender, somewhat flattened, slightly tapered towards the posterior extremity.

Branchial fans large, about one-third the length of the body; of a dark purple colour towards the base, where the filaments are all united by a web for a short distance. Each filament is marked on the smooth rachis at regular distances with two small round purple spots. There are five pairs of these spots on each filament, the first being near the base and the last a short distance from the apex. The filaments are all rather densely and closely ciliated on one side. The two tentacula are smooth and setaceous, short, stout, and sharp pointed at the apex. The collar is slightly lobed; and the upper part of the body, on the ventral surface, a little below the head, is stained with a rather broad dark purple mark, and along each side of the body, at the base of each foot, is a small spot of the same colour. The thorax has ten pairs of setigerous feet, and the purple spots at their bases are much larger than those at the base of the feet belonging to the abdominal segments. In one specimen (which I cannot see differs specifically from the others in other respects) the setigerous feet are twelve pairs in number and the body is somewhat broader.

The tube which the animal constructs, and in which it lives, is narrow, about the circumference of a swan-quill, long, round, and consists of a toughish membrane covered with a rather thick, smooth coat of mud.

The length of the animal is about 3 inches; that of the tube about $4\frac{1}{2}$ inches.

Hab. Island of St. Thomas, West Indies. Collected by M. Sallé. (Mus. Brit.)

2. SABELLA NIGRO-MACULATA, *Baird.* Pl. V. figs. 5 & 6.

Worm rather short, broad and stout, tapering slightly near the posterior extremity (fig. 5).

The branchial fan consists of numerous short filaments united near the base by a web, and about the fourth part of the length of the body. They are of a dark brown colour spotted with white on the rachis, are densely ciliated on one side with long stout cilia, and on the rachis, which is smooth, there are at regular distances about twenty other very short filaments, set in pairs (fig. 6). Near the base of the filament spring a pair longer and broader, and near the middle of its length another pair of the same kind. The collar is deeply lobed and of a dark purple colour. The body throughout its whole length is spotted with numerous dark purple or nearly black dots of various sizes, but largest on the superior extremity. The thorax possesses seven pairs of setigerous feet. The two smooth filaments are short and flat, and sharp pointed at the apex. The tube in which the animal lives is rounded, and is composed of a toughish membrane covered outwardly with a smooth coat of mud.

The length of the animal is about $2\frac{1}{2}$ inches; that of the tube nearly double the length.

Hab. Island of St. Vincent, West Indies. From the Rev. Lansdowne Guilding's Collection. (Mus. Brit.)

3. SABELLA OCCIDENTALIS, *Baird.* Pl. V. figs. 7 & 8.

Worm slender, of a cylindrical form, slightly tapering towards the posterior extremity (fig. 7).

Branchial fan composed of about sixteen filaments on each side. The filaments are densely ciliated on one side; the cilia of a yellow colour for most part, interspersed at short distances with black cilia, generally disposed in pairs, or in clusters of three, and rather stouter than the others. The rachis is smooth, but dotted along one side with numerous very small black spots (fig. 8). The filaments are all united near their base by a web, which is of a dark purple colour. The two smooth filaments are short, and sharp pointed at the apex. The collar is narrow and slightly bilobed. The thorax has seven pairs of setigerous feet. We possess no tubes belonging to the specimens.

Hab. Island of St. Vincent, West Indies. From the Collection of the Rev. Lansdowne Guilding. (Mus. Brit.)

4. Sabella grossa, *Baird*.

Worm remarkably thick, short and solid-looking, of a uniform dark olive colour, and about the same dimensions anteriorly as posteriorly.

The branchial filaments of the only specimen we possess have unfortunately been destroyed, but the peduncle upon which they were placed remains, and exhibits a spiral twist like that represented in M. Milne-Edwards's figure of *Sabella unispira* in Cuvier's 'Animal Kingdom' (Crochard edition, t. 4. fig. 1 *a*). The collar is everted, thick, and bilobed. The thorax possesses eight pairs of setigerous feet. The smooth (?) tentacles are wanting; the specimen, which has been for a good many years in the Museum Collection, being in only tolerable preservation. There is no tube belonging to the specimen. Length of animal (without branchiæ) about $4\frac{1}{2}$ inches, breadth about 7 or 8 lines. In general appearance it resembles *S. melania* of Schmarda from Jamaica.

Hab. Island of St. Helena. From the Collection made during the voyage of H.M.S. 'Chanticleer.' (Mus. Brit.)

5. Sabella grandis, *Baird*.

Worm of a rather square or quadrilateral shape, tapering slightly to the extremity, which terminates in a sharp point. Length (without branchiæ, which unfortunately have been lost) about $6\frac{1}{2}$ inches.

Collar rather broad and deeply bilobed. Thoracic feet seven pairs. Segments belonging to them smooth, not grooved on the upper dorsal surface. Body of a dark brown colour on the back, rather yellow underneath or on the ventral surface. Feet numerous, about 100 in number. Peduncles large, well developed. Anterior and posterior divisions separated by a groove, in the centre of which are situated the feet. Along the dorsal surface, with the exception of the seven first or thoracic segments, there runs a deep groove dividing each segment into two halves. The setæ of the feet appear in many instances to be enveloped in a membranous little bag, which, falling off, allows the setæ to project. These are short, slender, smooth, setaceous, and very sharp pointed.

The case in which the worm lives is somewhat larger than the animal itself, and is a round and leathery-looking tube covered over externally with a thin coat of mud.

Hab. Coast of New Zealand. From the Collection of Sir A. Smith, M.D. (Mus. Brit.)

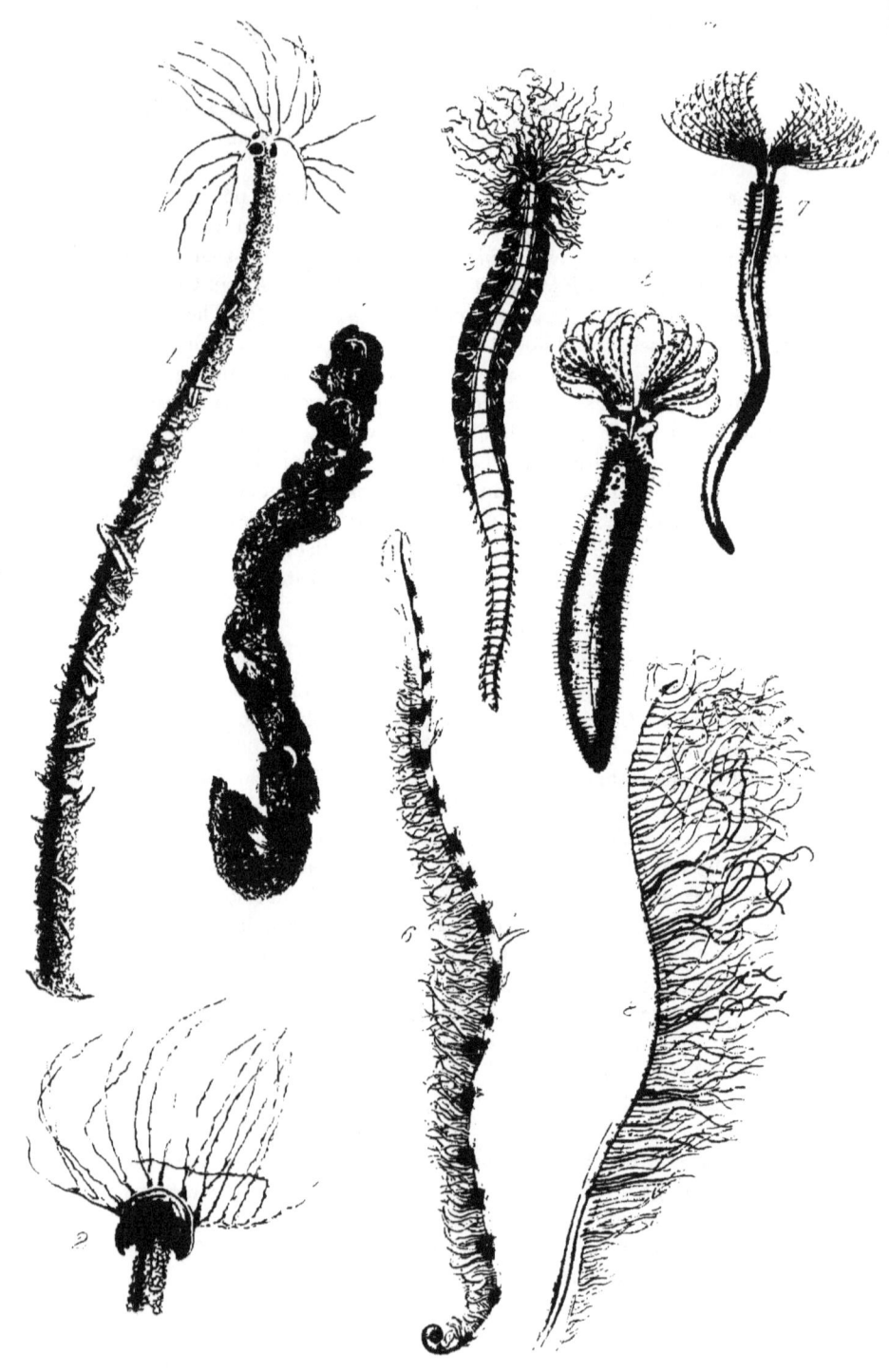

With the Author's Comp

margin. Beneath bronzy; apical segment of abdomen and legs ferruginous.

Head deeply and rugosely punctured. Thorax not quite two-thirds as long as wide at the base; anterior margin slightly produced in the centre, about half as long as the base; sides gradually but slightly rounded; base with a yellow median lobe; surface deeply and largely punctured, especially at the sides; dorsal line, and a line parallel to each lateral margin, impressed. Elytra deeply punctate-striate, twice as long as wide; posterior margins finely and sharply denticulate; apex of each bidentate. Beneath and legs punctured, covered with long hairs.

Length $9\frac{1}{2}$ lines. Breadth $3\frac{1}{4}$ lines.

Hab. South America.

Remarks on several Genera of *Annelides*, belonging to the Group Eunicea, with a notice of such Species as are contained in the Collection of the British Museum, and a description of some others hitherto undescribed. By W. BAIRD, M.D., F.R.S., F.L.S., &c.

[Read February 4, 1869.]

THIS group of *Annelides* contains individuals remarkable for their great length; and, according to M. Quatrefages, some of the species exhibit a complication of structure superior to that of any other of the *Annelida errantia*. They are of an elongated form, and generally slender, and are composed of numerous articulations. The head is more or less distinctly lobed, and possesses from five to seven organs usually described as antennæ or tentacles, and sometimes two tentacular cirri on the dorsal part of the buccal segment. The feet are disposed in one row only; and the branchiæ, which are present in all, are either pectinated and occasionally much developed, or simple and small. The mouth is armed with strong denticulated maxillæ.

This group may be divided into two families, EUNICIDÆ and ONUPHIDIDÆ.

Family I. EUNICIDÆ.

The species of this family are characterized by the head being distinctly two- or four-lobed, by the presence of two eyes, and the possession of five long and filiform organs generally described as antennæ or tentacles. One of these is single, placed in the centre, and, following the terminology adopted to distinguish these organs in the Aphroditacea, may be described as the *tentacle*; two are intermediate, the *antennæ*; and two external, the *palpi*. Feet armed with simple and compound setæ; usually with one or two spines (aciculæ) and one or two forcipate setæ or hooklets (uncini). Branchiæ generally pectinated and well developed.

To this family belong only two genera, *Eunice* and *Marphysa*.

Genus I. EUNICE.

Head two- or four-lobed. Buccal segment carrying on its dorsal side two rather short tentacular cirri of the same structure as the tentacle and antennæ. Simple setæ, in most of the species, of two kinds:—one long, flagelliform, sometimes lanceolate or limbate, and always acutely pointed (*simple*); the other slender, but terminating in a broad head, which is surmounted by a series of small teeth, like the teeth of a comb (*pectinate setæ*). The compound setæ have the appendage short, falciform, and toothed on its internal edge. The spines (*aciculæ*) are generally stout, dark-coloured, and obtuse. The hooklets (*uncini*) are, in the greater number of instances, terminated, as it were, with two teeth like those of a forceps (*forcipate*).

The species are rather numerous, forty-five having been enumerated by M. Quatrefages. Of these, unfortunately, the Museum possesses only a few. Several new species, however, occur, which, after a short notice of those contained in our collection, I shall briefly describe. I shall arrange those which I have been able to examine, according to the different appearance of the uncini or hooklets of the feet. Perhaps they may be worthy of separate generic denominations.

I. Species in which no uncini or hooklets are present.

This division is perhaps equivalent to the genus *Eunice* as restricted by Malmgren.

II. Species in which uncini are present, but are not forcipate at the apex.

III. Species which possess uncini forcipate at the apex.

These two divisions may be equal to the genus *Leodoce* of Savigny as restricted by Malmgren.

N.B. All the species mentioned here are in the British Museum.

I. *Species which do not possess uncini.* ?=Eunice *restricted.*

Sp. 1. EUNICE APHRODITOIS.

Nereis aphroditois, *Pallas*, 1788.
Terebella aphroditois, *Gmelin*, 1789.
Eunice gigantea, *Cuvier*, 1817?, *Grube, Quatrefages*, &c.
Nereidonta aphroditois, *Blainville.*
Leodoce gigantea, *Savigny.*

There appears to be some confusion with regard to the synonymy of this species. Quatrefages adopts the name of *gigantea*, and quotes, as the type of it, the *Nereis gigantea* of Linnæus. Referring, however, to the ' Systema Naturæ,' we find Linnæus quoting, as the type of his species, the *Millepoda marina Amboinensis* of Seba, 'Thesaurus,' tab. 81. fig. 7, which, as Savigny has already shown, and which, as I have mentioned in my previous paper on the *Amphinomacea* (vide 'Proceedings of Linnean Society' for 1868, vol. x. p. 219), is in reality the *Amphinome carunculata* of Pallas.

Cuvier, in his first edition of the 'Règne Animal,' named the present species *Eunice gigantea*; but Pallas had long anteriorly described and figured it under the denomination of *Nereis aphroditois*. Quatrefages describes a new species under the name of *Eunice Roussæi*; but this I consider to be identical with the *aphroditois*. He quotes Cuvier's *gigantea* for both; and indeed it would appear that his chief reason for making *two* species is the difference of their habitat, the one being a native of the Atlantic Ocean and the West Indies, whilst the other is from the Indian Seas, Isle of France, &c. We have a variety of specimens of what I consider to be the true *aphroditois*, from Australia and Van Diemen's Land, so that in all probability this species is to be found in various parts of the world.

The head-lobes in all our specimens are four in number. Quatrefages says of his species *E. gigantea=aphroditois*, " Caput quasi sex-lobatum." Savigny expressly says of his *Leodoce gigantea*, "tête à quatre lobes."

This is one of the longest of known *Annelides*, one specimen we possess in the British-Museum collection being 41 inches, or nearly 3½ feet long.

Hab. Van Diemen's Land, Freemantle, W. Australia, New Holland, *Mus. Brit.*; Indian Seas, Isle of France, *Quatrefages* (*gigantea*); Atlantic Ocean, West Indies, *Quatrefages* (*Roussæi*).

Sp. 2. EUNICE ELSYI, *Baird*.

Body about 4½ inches long, and consisting of about 120 segments. Head with two lobes, which are round and very prominent. Tentacle, antennæ, and palpi moniliform, of considerable length. Tentacle longer than antennæ, being equal in this respect to the transverse diameter of the first seven segments. Tentacular cirri moniliform also, and about equal in length to the transverse breadth of the buccal segment. This segment is equal to the transverse breadth of the four succeeding segments, and has its ventral margin not crenated, but cleft by two short incisions in the centre. Branchiæ commencing on the sixth foot; pectinations or branchlets at first only three or four, increasing in number in the succeeding segments to eighteen. Dorsal cirri moniliform, rather long and finely pointed. Ventral cirri short and conical. Anal cirri moniliform, of moderate length.

Feet:—Simple setæ long, filiform, and acutely pointed. Pectinate setæ, with numerous fine pectinations or teeth, the outer one at both sides being a little longer than the others. The compound setæ have the shaft rather short and stout, and the falciform appendage bluntly toothed, the teeth being rather obtuse. Aciculæ two, strong, dark-coloured, and bluntly pointed. No uncini or hooklets visible.

Hab. North Australia, *Elsey*.

Sp. 3. EUNICE MADEIRENSIS, *Baird*.

=? Eunice adriatica, *Schmarda, Neue wirbellose Thiere*, i. p. 124, tab. 32. fig. 257.

Body convex dorsally, flat ventrally, with a furrow running down the centre, wrinkled throughout. Our specimens are imperfect at the caudal extremity; but they consist of about 286 segments, and are about 5 inches long. Head with two lobes. Buccal segment and portion to which the tentacular cirri are attached equal in length to the transverse diameter of the four succeeding segments; its ventral margin smooth, not crenated, and straight-edged. Tentacle, antennæ, and palpi rather short. Tentacle longer than antennæ, indistinctly moniliform. Tentacular cirri short, not jointed, and not equal to the transverse diameter of the buccal segment. Dorsal cirri rather long and slender, ventral cirri short and conical. Branchiæ not commencing till near the 200th segment, and consisting of only one moderately long filament.

Feet:—Simple setæ long, lanceolate, finely toothed or serrated on the inner margins and acutely pointed. Compound setæ long, but rather shorter than the simple setæ; falciform appendages with two small teeth, one a little below the apex, the other a little lower down. Spines or aciculæ three in number, all straight and blunt-pointed. Neither pectinate setæ nor uncini were visible.

Hab. Madeira.

This species approaches so closely to the *E. adriatica* of Schmarda that I can scarcely separate the two. The only marked differences are the structure of the compound setæ and the habitat. In *adriatica* the falciform appendage is, as it were, forcipate or bifid at the apex, whereas in this species (*madeirensis*) it is as in most of the other known species, bidentate, one tooth just below the apex, the other lower down. In both species the pectinate setæ and uncini appear to be absent.

II. *Species in which the uncini are only curved at the extremity, not forcipate or hooked.* ?=Leodoce, *Savigny, as restricted.*

Sp. 4. EUNICE NORVEGICA.

Nereis norvegica, *Linnæus, Syst. Nat.* 12th edit. p. 1086.
Nereis pennata, *Müller, Zool. Dan.* i. 30, tab. 29. figs. 4–7.
Nereis pinnata, *Müller, l. c.* tab. 29. figs. 1–3.
Eunice norvegica, *Cuvier, Règn. Anim.* iii. 100; *Aud. & M.-Edwards; Grube et auctor. var.*
Leodoce norwegica, *Savigny, Syst. des Annélides,* p. 51; *Lamk. An. s. Vert.* 2nd edit. tom. v. p. 562.
Leodoce norvegica, *Malmgren, Ann. Polychæta Spetsberg. &c.* p. 64.
Nereidonta norvegica, *Blainville, Dict. des Sc. Nat.* art. Vers.
Nereidonta pinnata, *Blainville, l. c.*
Eunice norwegica, *Quatrefages, Hist. des Annelés,* i. 324.
Eunice pinnata, *Quatrefages, l. c.* 325.

In this species the simple setæ, the pectinate setæ, and the compound setæ are present. The spines or aciculæ are two in number; and there is only one uncinus or hooklet. This is shorter than the spines, more slender, more sharply pointed, and curved but not forcipate at the apex.

Hab. Our specimens are from Bohuslän, Sweden.

Sp. 5. EUNICE TENTACULATA.

Eunice tentaculata, *Valenc. MS.*; *Quatrefages, Hist. Ann.* i. 317.
Not Eunice tentaculata, *Kinberg, Fregatt. Eugen. Resa,* tab. 15. f. 13.

Hab. Van Diemen's Land (*Mus. Brit.*); Port Western, *Quatrefages.*

In this species simple setæ, pectinate setæ, and compound setæ are present. The spines or aciculæ are two; but there is only one uncinus, which is similar in form to the spines, is strongly curved and not forcipate at the apex.

III. *Species in which the uncini are forcipate at the apex.*
=? Leodoce, *Savigny, as restricted.*

Sp. 6. EUNICE ANTENNATA.
Leodice antennata, *Savigny, Syst. des Annélides,* p. 50.
Eunice antennata, *Cuvier, Audouin & M.-Edwards, Grube, &c.*
Hab. Cosseir, Red Sea.

Sp. 6*. EUNICE ANNULICORNIS.
Eunice annulicornis, *Johnston, Cat. of Non-parasitical Worms,* p. 131.
Leodoce annulicornis, *Spinola, MS.?*

This species was described by Dr. Johnston from a specimen contained in the collection of the British Museum. It was named *Leodoce annulicornis* in our collection—a specific name which Johnston adopted. The label was marked "Spinola," and it was erroneously considered by Dr. J. that that name was the habitat whence it came. It is in reality the *E. annulicornis* of Maximilian Spinola, but, I believe, only a MS. name; and its native habitat may probably be the Gulf of Genoa. The simple setæ are long, lanceolate flagelliform, and long and acutely pointed. Pectinate setæ long, slender, the outermost tooth of the pectinated head being much elongated beyond the others and straight. Compound setæ with the shaft stout, broadly triangular at the apex, where the appendage is fixed, and striated; falciform appendage bidentate, teeth rather small. Aciculæ two or three; on the upper feet there appear to be three, and no uncini. On the lower feet there appear to be only one acicula and one uncinus, which is curved in its length, and has the apex merely emarginate and not forcipate.

In this species the three kinds of setæ are present—simple, pectinate, and compound. The spines or aciculæ are two in number; but there is only one uncinus, nearly equal in size to the aciculæ, and forcipate at the apex.

Sp. 7. EUNICE HARASSII.
Eunice Harassii, *Audouin & M.-Edwards, Litt. de la France,* ii. 151, tab. 3. figs. 5–7, 10, & 11; *Cuvier, Grube, Quatrefages, &c.*
Hab. Southern shores of England (*Mus. Brit.*); Coast of Normandy, &c., *Quatrefages.*

There is only one uncinus, curved and forcipate at the apex. The setæ and spines are all present as in the preceding species.

Sp. 8. EUNICE MACROCHÆTA?

Eunice macrochæta? *Schmarda, Neue wirb. Thiere*, i. 128, fig. xylogr. app.

Hab. In holes of coral rocks in Jamaica.

In this species, which I consider to be identical with the *Eunice macrochæta* of Schmarda, the three kinds of setæ, simple, pectinate, and compound, are present. There is only one spine, which is strong, straight, and club-shaped at the apex, and one uncinus, which is strongly forcipate.

Sp. 9. EUNICE QUOYA?

? Eunice Quoya, *Valenc. MS., Quatrefages, l. c.* p. 318.

Hab. North Australia, *Elsey.*

The specimen we possess is in such bad condition that I can only refer it with doubt to the species described by Quatrefages. The falciform appendage of the compound setæ is slender and destitute of teeth. There is only one spine, but two uncini, which are smaller and lighter-coloured than the spine, curved and forcipate at the apex.

Sp. 10. EUNICE FIJIENSIS, *Baird.*

Body slender, segments about ninety-eight in number. Branchiæ commence upon the seventh foot-bearing segment. The setæ of the feet are not numerous. The simple setæ are rather broadly lanceolate and very sharp-pointed. Pectinate setæ appear to be absent altogether. The compound setæ have the falciform appendage bidentate at the apex. There are two spines or aciculæ, which are stout, swollen in the middle of their length, and slightly curved at the point. Only one uncinus or hooklet is present, which is strongly and distinctly tridentate at the apex, and has the shaft curved. It approaches somewhat to the *Eunice gracilis* of Grube, from Tahiti.

Hab. Fiji Islands (*Mus. Brit.*).

Sp. 11. EUNICE WOODWARDI, *Baird.*

=? Leodoce hispanica, *Savigny, Syst. des Annél.* p. 51.

Body cylindrical, smooth, of a light iridescent colour, about $1\frac{1}{2}$ inch long. Head with two lobes. Buccal segment rather narrow, not much broader than the following segment. Tentacle, antennæ, and palpi rather long, ringed with black marks, but not jointed or moniliform. Tentacle longer than antennæ. Tentacular cirri short, a little longer than the transverse diameter of the buccal segment. Dorsal cirri rather long. Ventral cirri short and conical. Branchiæ commencing about the third segment; pectinations filiform.

Feet:—Simple setæ long, lanceolate, acutely pointed, and finely toothed or serrated on the inner edge or margin for a part of their length. Pectinate setæ small, apparently few in number, and with few pectinations or teeth. Compound setæ short, about half the length of the others; falciform appendage with a sharp tooth just beneath the apex, and a blunter one nearer the lower portion. Aciculæ or spines two, slightly curved, dark-coloured and blunt-pointed. Uncini or booklets several in number, but varying from two to five, curved and tridentate at the apex, lying across the aciculæ.

Hab. Corunna, *H. Woodward.*

Sp. 12. EUNICE ANTARCTICA, *Baird.*
?=Eunice havaica, *Kinberg, Fregatt. Eugen. Resa,* tab. 15. figs. 14 *b-g.*

Body slender, of a dark æneous colour, and consisting of from 115 to 120 segments. Buccal segment scarcely equal to the two succeeding ones. Head with two lobes. Tentacle, antennæ, and palpi articulated. Tentacle longer than antennæ. Tentacular cirri longer than the transverse diameter of the buccal segment, and articulated. The lobe or segment from which they spring is of about the same breadth as the succeeding segment. Branchiæ small, commencing about the eighth pair of feet, and terminating about the thirty-eighth segment. Anal cirri of considerable length, indistinctly articulated. Dorsal cirri slender. Ventral cirri stout, conical, not so long as the dorsal, but much stronger.

Feet rather small. Simple setæ long, flagelliform and sharp-pointed. Pectinate setæ few in number, rather small, with the outer tooth longer and stronger than the others. Compound setæ short; falciform appendage small, with a small sharp tooth a little below the apex. Spines or aciculæ two, slightly curved and obtusely pointed. Uncini or booklets two, curved, and forcipate at apex.

Hab. Antarctic Seas, *Antarctic Expedition.*

Sp. 13. EUNICE PLICATA, *Baird.*

Body cylindrical, tapering towards the inferior extremity, from 2 to 3 inches long, and consisting of about 130 short or narrow articulations. Buccal segment nearly equal to the four succeeding articulations, with the ventral margin prominent, standing high up, and separated as it were from the upper lobe, which is not crenated but plicated on both upper and lower margins with numerous small plaits running down the ventral side of the segment. Tentacle, antennæ, and palpi indistinctly articulated, rather short, and ringed at intervals with dark bands. Tentacular cirri not equal in length to the transverse diameter of the buccal segment. The branchiæ commence about the seventh segment; pectinations few throughout, about five in number. Dorsal cirri stout but not long. Ventral cirri short and conical. Anal cirri ringed with dark rings.

Feet:—Simple setæ numerous, lanceolate, very long and very finely pointed. Pectinate setæ few in number; pectinations or teeth numerous, and, as it were, double, one row beneath another, the outermost tooth on each side being the longest and strongest. Compound setæ shorter than the simple setæ; falciform appendage bidentate; one tooth sharp and prominent, a little below the apex, the other blunt and near the lower part. Spines or aciculæ appear to be three in number; two dark-coloured, stout, straight, blunt-pointed, and rather long; a third shorter, lighter-coloured, and obtuse at extremity. One uncinus only, of considerable length, curved, and forcipate at the apex.

Hab. Freemantle, Australia, *Dr. Bowerbank.*

Sp. 14. EUNICE BOWERBANKI, *Baird.*

Body stout, tapering towards the lower extremity, and consisting of about 140 articulations. Buccal segment broad, nearly equal in length to the transverse diameter of the first three segments of the body. Whole body of a metallic lustre. Tentacle, antennæ, and palpi rather short, moniliform. Tentacle and antennæ of nearly equal length. Tentacular cirri articulated, equal in length to the transverse diameter of the buccal segment. Ventral margin of the buccal segment not crenate. The branchiæ commence on the fifth segment, quickly arrive at the maximum number of pectinations, soon decrease in size, but continue to be present till near the extremity. Dorsal cirri short, stout, articulated. Ventral cirri short and stout.

Feet:—Setæ unusually long. Simple setæ long and very sharp-pointed. Pectinate setæ with rather numerous pectinations, the outside tooth prolonged and slightly curved inwards. The compound setæ have the falciform appendage strongly bidentate; one tooth (as usual) under the apex, the other on the lower portion.

The spines or aciculæ are two, long, straight, or only slightly curved at the apex, one being shorter than the other. There is only one uncinus or hooklet, which is shorter than the spines, curved, club-shaped at apex, and indistinctly forcipate.

Hab. Australia, *Dr. Bowerbank.*

Sp. 15. EUNICE GUTTATA, *Baird.*

Body broad, flat, except near the anterior extremity, which is somewhat cylindrical, very gradually tapering towards the tail, and consisting of about 120 very narrow segments. Length nearly two inches, breadth about the centre of the body $2\frac{1}{2}$ lines. Along the lower portion of the body the back is marked with several large dark spots. Tentacle, antennæ, and palpi not articulated, and moderately long. Tentacular cirri very short. Head indistinctly four-lobed. Buccal segment about equal to the breadth of the four succeeding segments. Ventral margin of buccal segment swollen and crenate. The branchiæ commence at the sixth segment, and are small and dark-coloured; pectinations about the centre of the body, ten in number. Dorsal cirri rather large.

Feet small. Simple setæ lanceolate and finely pointed. Pectinate setæ finely toothed, the external tooth longer than the others. Compound setæ stout, and broad at the summit of the shaft; falciform appendage rather stout, curved at the apex, and with only one tooth, which is a little below the apex. Spines or aciculæ three in number—two long and stout, dark-coloured, and blunt at the point, the third much smaller but of exactly the same form. Uncini or hooklets two, lighter-coloured than the spines, curved, and sharply forcipate.

Hab. Taken between Bombay and Singapore.

Sp. 16. EUNICE NARCONI, *Baird.*

Body slender, nearly 2 inches long, and consisting of about 120 segments. Head with two prominent lobes. Buccal segment about equal to the three succeeding, the articulation from which the tentacular cirri spring being of itself equal in size to the first segment of body. Ventral margin of buccal segment not crenated. Tentacle, antennæ, and palpi inconspicuously articulated. Tentacle longer than the antennæ and palpi, which are all of about equal length. Tentacular cirri longer than the transverse diameter of the buccal segment. Dorsal cirri of moderate length, conical. Ventral cirri short. Branchiæ very small, commencing on the twelfth segment.

Feet:—Simple setæ long, fine and acutely pointed. Pectinate setæ few in number and small. Compound setæ only half the length of the simple setæ; falciform appendage small, with only one small tooth under the apex. There are two spines or aciculæ, light-coloured and slightly curved at the apex, which is obtuse, and only one uncinus or hooklet, which is light-coloured also, curved, and forcipate at the apex.

Hab. Island of Narcon, Antarctic Seas, *Antarctic Expedition.*

Sp. 17. EUNICE GUILDINGI, *Baird.*

Body about $5\frac{1}{2}$ inches long, tapering very much towards the tail, which portion of the body is quite cylindrical. Convex dorsally, flattish ventrally for about half its length. The anterior portion of the body is about 3 lines in breadth, and the posterior only about 1. Body of a dark rufous colour, with very little iridescence. Head with two lobes. Buccal segment about equal to the two next succeeding. Ventral margin of buccal segment not crenate. Tentacle, antennæ, and palpi short, rather thick, and indistinctly moniliform; tentacle a little longer than antennæ. Tentacular cirri about equal to the breadth of the buccal segment. Dorsal cirri very indistinctly articulated, of moderate length. Ventral cirri very short. Branchiæ commencing at about the fifth pair of feet, rapidly attaining their greatest development, but quickly afterwards diminishing in number of pectinations, and at about half the length of the body disappearing altogether.

Feet:—Simple setæ long, lanceolate, finely denticulate or serrate at one side for half their length, and acutely pointed. Pectinate setæ broad at the apex, but the pectinations or teeth not very distinct, outermost one most prominent; these setæ are of unequal size, some being much smaller than the others. Compound setæ with the falciform appendage strongly and sharply bidentate, one tooth a little below the apex, the other nearer the lower portion. There is only one spine, which is straight, dark-coloured, and obtusely pointed, and only one uncinus or hooklet, which is curved and indistinctly forcipate at apex.

Hab. St. Vincent's, West Indies, *Guilding.*

Genus II. MARPHYSA.

Leodocæ marphysæ, *Savigny, l. c., Grube, &c.*
Eunice (sp.), *Cuvier, Audouin & M.-Edwards, &c.*
Marphysa, *Quatrefages.*

Head with or without lobes. No tentacular cirri on back of

buccal segment. Eyes, tentacle, antennæ, palpi, and branchiæ as in *Eunice*. Compound setæ having the falciform appendage frequently long and without teeth on its edge.

Sp. 1. MARPHYSA SANGUINEA.
   Nereis sanguinea, *Montagu, Linn. Trans.* xi. 20. t. 3. f. 1.
   Leodoce opalina, *Savigny, l. c.*
   Nereidonta sanguinea, *Blainville, Dict. Sc. Nat.*
   Eunice sanguinea, *Cuvier, Audouin & M.-Edwards, Grube, &c.*
   Marphysa sanguinea, *Quatrefages, l. c.*

In this species, which has been the object of much and careful investigation by M. Quatrefages, and which is not uncommon on our southern coasts, the simple setæ of the feet are long, narrowly lanceolate, with a much elongated and acute point. The pectinate setæ are of two kinds :—one slender, broad at the apex, finely pectinate, like the teeth of a small-tooth comb, and having the outermost longer and stronger than the others; the other coarser, shorter, and having the teeth much stronger and fewer in number, like those of a large-tooth comb. Compound setæ rather slender, with the falciform appendage long, slender, and terminating in a sharp point, which is straight, and not toothed on the edge. The aciculæ are four or five in number in the upper feet, three in those of the middle third, and only two in the feet of the lower third of the body: one of these is generally smaller than the others and of a lighter colour, as if it were taking the place of the uncini; but, like the aciculæ, they are nearly straight, obtuse at the point, and not forcipate. The pectinate setæ, of both kinds, are much more numerous in the feet of the lower third of the body; and the compound setæ become fewer and more slender.

Hab. South coast of England, Falmouth, South Devon, and Polperro, coast of Cornwall (*Mus. Brit.*); coast of France, *Quatrefages*.

In our own collection at the British Museum we have from Polperro a small Annelid which is either a young one or a small variety of the *M. sanguinea*. The only differences I can observe are, its being much more slender in all its proportions, and the aciculæ being only three in the feet of the upper third, and only one in the middle and lower third of the body, while there is a distinct uncinus, somewhat curved and slightly forcipate at the apex.

Sp. 2. MARPHYSA PARISHII, *Baird.*
Body about 32 inches long, composed of about 350 segments,

of a dark æneous colour throughout, and iridescent. Head small, with two comparatively large lobes. Labrum crenulated. Buccal segment equal in size to the three succeeding ones. Tentacle, antennæ, and palpi annulated, short, not much longer than just to reach beyond the margin of the head; all of them of about equal length. Body narrower at the two extremities; anteriorly it is so only for a short distance, about the first seven segments, then becoming broad for a short distance, and again contracting as it descends posteriorly. The first six or eight of the anterior segments are wide, then they become narrow, and the breadth of the body becomes greater till about the 60th segment, when they again begin to increase in width, while the body itself begins to decrease in breadth. This continues till they approach the tail, when the thirty or forty last segments again become narrower. The tail appears to be destitute of caudal cirri. The branchiæ are pectinate, though they do not appear to exceed five or six filaments in number; they commence about the 24th segment, and continue till near the lower extremity.

Feet:—On the 24th pair the setæ are of two kinds only; they are numerous and long. The simple setæ are curved and flagelliform, or elongate-lanceolate. The compound setæ are slender; the falciform appendage is long and slender, finely pointed, without teeth, having exactly the appearance of a long slender bayonet. There are four strong and black-coloured aciculæ, blunt-pointed, and a fifth not quite so strong or long as the others (? takes the place of the uncinus). About the 60th foot or thereabouts there appear to be only the four aciculæ; and two of these are not so obtuse at the point as the others; the setæ are exactly similar to those of the anterior feet. In the feet of the lower portion of the middle third of the body, the simple setæ are precisely the same as those above: the aciculæ are only two, dark and blunt-pointed; but two others accompany them, of a much lighter colour, not so strong, and slightly curved at the apex. Pectinate setæ appear now; they are of two kinds, one slender, with the head broad and a number of very fine teeth, the other coarser, the head slightly oblique, and the teeth strong and reduced to the number of from four to six, resembling very much in appearance a five-short-pronged silver fork.

Hab. Brazil, *Capt. John Parish, R.N.*

N.B. These are the only species of this genus the British Museum at present possesses. The described species are not very numerous, Quatrefages enumerating only nine.

Family II. ONUPHIDIDÆ.

Onuphididæ, *Malmgren, Annulat. Polychæta Spetsberg. &c.* 1867.

Head with lobes as in Eunicidæ, furnished with seven organs usually described as antennæ or tentacula. Two spring from the front of the head, and are very short (*antennules*). The remaining five are as in Eunicidæ. Tentacular cirri generally present, sometimes wanting. Branchiæ pectinated or plumose—or simple, consisting of only one filament. Eyes two. Maxillæ as in Eunicidæ.

To this family may be referred five genera, four of which have already been described, which may be thus characterized:—

I. ONUPHIS, *Audouin & M.-Edwards.* Branchiæ pectinate. Tentacular cirri placed laterally on buccal segment. Tentacle, antennæ, and palpi annulated more or less throughout their whole length.

II. DIOPATRA, *Audouin & M.-Edwards.* Branchiæ plumose, the branchlets disposed in a spiral tuft round a central stalk. Tentacular cirri placed on the dorsal portion of the buccal segment. Tentacle, antennæ, and palpi strongly annulated at the inferior or basal portion only.

III. TRADOPIA (*gen. nov.*). Branchiæ pectinate. Tentacular cirri placed on dorsal part of buccal segment. Tentacle, antennæ, and palpi strongly annulated on the lower or basal portion.

IV. NOTHRIA, *Johnston* (*Malmgren*). Branchiæ bipartite only. Tentacular cirri placed on the dorsal portion of buccal segment. Tentacle, antennæ, and palpi simple, not annulated on any portion of their length.

V. HYALINŒCIA, *Malmgren.* Branchiæ reduced to a simple branchial filament. No tentacular cirri. Tentacle, antennæ, and palpi annulated at the inferior or basal portion.

Genus I. ONUPHIS, *Audouin & M.-Edwards, Litt. de la France,* ii. 151, t. 3A. figs. 1–5; *Malmgren.*

We have no specimens belonging to this genus in the British-Museum collection.

Genus II. DIOPATRA, *Audouin & M.-Edwards, l. c.* 155; *Kinberg, Malmgren, &c.*

Branchiæ plumose, not pectinate, but consisting of a tuft of many short filaments rolled in a spiral form round a central stalk.

Tentacular cirri of moderate length, arising from under the dorsal edge of buccal segment. Tentacle, antennæ, and palpi strongly annulated on the lower portion or root, which is of considerable length.

Sp. 1. DIOPATRA AMBOINENSIS.

Diopatra amboinensis, *Audouin & M.-Edwards, Litt. de la France,* ii. 156, tab. 3 A. figs. 6–8; *Grube, Quatrefages, &c.*

In this species, of which we possess only one specimen, the simple setæ of the feet are of two kinds—one shorter than the other, slightly elbowed near the point, which is very acute, the other much longer, lanceolate, and indistinctly or very minutely serrate on the margins. Pectinate setæ numerous, the broad head on which the teeth are placed, with its pectinations, being disposed obliquely. There appear to be two uncini or hooklets to each foot, of a stout form, and forcipate at extremity, but no aciculæ.

Hab. Amboina, *M.-Edwards.* (No habitat to our specimen.)

Genus III. TRADOPIA, *Baird.*

Branchiæ pectinated. Tentacular cirri placed on dorsal part of buccal segment. Tentacle, antennæ, and palpi strongly annulated on inferior or basal portion, which is more than half the length of the whole organ.

Sp. 1. TRADOPIA MACULATA, *Baird.*

Body about 8 inches long, narrow, tapering gradually to the tail. Rather flat on dorsal surface, and somewhat convex anteriorly on the ventral surface. Head rather small. Antennules very short. Tentacle, antennæ, and palpi very different from each other in length: tentacle and antennæ short, of about equal length; palpi long, nearly double the length of the antennæ. These organs are all finely and closely annulated on the basal portion, which is marked anteriorly with a row of dark spots throughout its whole length, are rather thick, and at least double the length of the anterior or terminal portion, which is short, smooth, and filiform. The tentacular cirri are short and slender, placed on dorsal portion of the buccal segment. The anterior feet are prominent and project upwards. The branchiæ are pectinate, commencing on the first foot, at first consisting only of two branchlets or filaments, but after the third or fourth becoming more pectinate. The tail terminates in two rather short cirri.

Anterior feet with the fascicle of bristles all simple, lanceolate, some, however, nearly double the length of the others; and in these feet no forcipate uncini, nor pectinate setæ, nor aciculæ are to be seen. In the feet, however, of the upper third of the body the pectinate setæ and the forcipate uncini make their appearance, the pectinate setæ with the head straight, and all the teeth or pectinations on the same plane, the forcipate uncini as in *Nothria* and *Onuphis*, while the simple setæ become more broadly lanceolate and limbate. The branchiæ on this part of the body consist of numerous branchlets or filaments, two stout branches at first being given off, which very shortly divide again into others, which again divide till there are about twenty filaments or branchlets. The main stems of these branchiæ are beautifully edged with blunt, round-pointed cirri; and all the branchlets or filaments appear as if beaded on their margins. The posterior feet retain the pectinate setæ and uncini; but the branchiæ are reduced (as in the first two or three feet) to two simple filaments.

Hab. Madras, *F. Day, Esq.*

Genus IV. NOTHRIA, *Johnston* (s. str. *Malmgren*).

Branchiæ bipartite. Tentacular cirri placed on the dorsal part of the buccal segment. Tentacle, antennæ, and palpi simple, not annulated on any portion of their length.

Sp. 1. NOTHRIA CONCHYLEGA.

Onuphis conchylega, *Sars, Beskr. og Jaktt.* p. 61, tab. 10. f. 28 *a–c*.
Onuphis Eschrichti, *Œrsted. Groenl. Ann. Dors.* 20, tab. 3. f. 33–41 & 45.
Diopatra Eschrichti, *Grube, Fam. Ann.* 43; *Quatrefages.*
Northia conchylega, *Johnston, Cat. Non.-Parasit. Worms*, 138.
Nothria *conchylega, *Malmgren, Ann. Polych. Spetsberg.* 66.

Hab. Coralline region, British coast, Berwick Bay (*Dr. Johnston*); Shetland Islands (*J. Gwyn Jeffreys*); South Devon (*J. Cranch*); North Seas (*Malmgren*).

Genus V. HYALINŒCIA, *Malmgren, Ann. Polych. Spetsberg.* 67.
Northia, *Johnston, Cat. Brit. Worms.*

Branchiæ reduced to one simple branchial filament. No ten-

---

* For some critical remarks on the genus *Nothria*, see at the end of genus *Hyalinœcia* following.

tacular cirri. Tentacle, antennæ, and palpi annulated at the inferior or basal portion.

Sp. 1. HYALINŒCIA TUBICOLA.
  Nereis tubicola, *Müller, Zool. Dan.* i. 18, tab. 18. f. 1-6.
  Leodoce tubicola, *Savigny, Syst. des Annél.* 383.
  Onuphis tubicola, *Sars, Beskr. og Jaktt.* 48; *Quatrefages.*
  Northia tubicola, *Johnston, Cat. Non.-Parasit. Worms,* 136.
  Hyalinœcia tubicola, *Malmgren, l. c.* 67.

Hab. Shores of Great Britain, Scotland, South Devon, Cornwall, &c. (*Brit. Mus.*); Asia Minor (*McAndrew*); North Sea (*Malmgren*).

As there seems to be some little obscurity about this species, I shall first describe the animal as it occurs in such specimens as we possess, and afterwards make a few critical remarks as to its name and position.

The body of the animal is generally about 2½ inches long. The tube which it inhabits, and which is horny, cylindrical, and exactly like the barrel of a small quill pen, is about 3½ inches long. The head presents the appearance, on the buccal surface, of two lobes, as in *Eunice*. The antennules are very short, rounded-oval, and are attached to the front of the apex of the cephalic segment. The tentacle is longer than the antennæ or palpi, and is annulated at the base. The antennæ, which spring from the head-lobe along with the tentacle, are of the same form as this organ, but a little shorter. The palpi spring from the side of the cephalic segment, and are still shorter than the antennæ. All these organs are annulated at the base, but do not appear to be jointed throughout their length. The eyes are two in number. The jaws are in three pairs—one pair curved, simple, a second strong and armed with twelve strong denticulations; the third pair are denticulated also, and armed with about six denticles. The two or three uppermost or most anterior of the feet are prominent, have a large setiferous tubercle with three cirri implanted on its surface, and project straight upwards. On the succeeding feet the ventral cirrus soon disappears, leaving only in its stead a round tubercle. The dorsal cirrus on the lower half of the body is long, lying on the back, and takes the place of a branchial organ of only one filament.

Feet:—Setæ of two kinds only, simple and pectinate. Simple setæ long, lanceolate, flattened or broadly limbate towards the upper half, and finely pointed. Pectinate setæ rather long; pectinations or teeth rather numerous, all on the same plane and equal in size; they vary in number, there being sometimes as many as ten in one fascicle. No compound setæ. Aciculæ or spines two to each foot on the middle or lower part of the body only, straight and very sharp-pointed. These spines are not round like those in the Eunicidæ, but are flat and more like simple setæ, taking the place of aciculæ. Uncini or hooklets generally two in number, a little shorter than the aciculæ, more cylindrical, and forcipate or bidentate at the apex.

Sp. 2. HYALINŒCIA BILINEATA, *Baird*.

Animal slender and narrow, gradually diminishing in size towards the tail, convex dorsally, and marked with two longitudinal reddish-coloured lines, which run throughout the whole length, one on each side. A small dark-red spot occurs between each foot. The organs attached to the head, antennules, tentacle, antennæ, and palpi, are very similar to those of *tubicola*; and the feet are furnished with only the same kinds of setæ as in that species. The simple setæ, however, are linear-lanceolate, not limbate or broadly lanceolate in the upper third as in *tubicola*. Dorsal and ventral cirri occur on the two upper thirds of the body.

These setæ and cirri vary considerably according to their situation. In the anterior pairs of feet there are no uncini or hooklets; but instead of them are two setæ very like the compound setæ of *Eunice*, only the falciform appendage (which is bidentate near the apex) is as it were soldered to the shaft and not moveable. Towards the middle of the body these compound-looking setæ disappear, and their place is taken by two regular forcipate uncini, as in *tubicola*. The shaft, however, is much curved, and it is by far the stoutest of all the setæ of the feet. On the lower portion of the body the simple setæ are of two kinds—one, three or four in number, being straight, stout, and very sharp-pointed, more like sharp-pointed aciculæ than setæ (are they aciculæ?), the others of the usual form, linear-lanceolate, about half the size of the others.

The cirri appear to be three in number on the segments of the upper third of the body. One of these is longer than the two others, and may be considered the branchial filament. About the middle third of the body this branchial filament disappears, only the two cirri being present. On the lower third of the body one of these cirri also disappears, one cirrus only remaining. Tail-cirri two in number.

The tube in which the worm lives is pellucid, soft, of a thin horny texture, and appears fitted closely to the body of the animal, wrapping it tightly around.

This species is much narrower and more slender than *tubicola*, and is altogether much smaller. The two longitudinal red lines running along its back are very distinct and characteristic. The cirri maintain the number of three for about a third of the length of the body, instead of only on three or four of the anterior feet; and the tube is very different from that of *tubicola*.

Hab. Off the coast of Cornwall, at a depth of from 20 to 40 fathoms water (*Laughrin, Mus. Brit.*).

Sp. 3. HYALINŒCIA VARIANS, *Baird*.

Worm about an inch in length, slender, of a slightly metallic lustre throughout. The tube is slender, cylindrical, about $1\frac{1}{2}$ inch long, of a horny substance, and contains the animal freely in it.

The head is rather small, but the organs springing from it are long. Antennules oval, springing from the anterior edge of the head, broader and considerably longer than those in *tubicola*. Tentacle longer than any of the other organs. Antennæ shorter than tentacle, and palpi shorter than antennæ. All these five organs are closely annulated at their base, having a distinct joint a little distance from the annulated portion, and being then indistinctly jointed at distant intervals during their length.

The anterior feet are rather prominent, and the cirri implanted upon them very short and small. The setiferous tubercle, giving origin to the fascicle of setæ, is placed between two large cirri in the succeeding pairs of feet. Dorsal cirri, or branchial filaments, on the upper half of the body long, then suddenly becoming shorter as they descend to the lower half. Setæ of two kinds only, simple and pectinate. Simple setæ long, very sharp-pointed, of a lanceolate form, and slightly curved, the flattened lanceolate portion being near the apex. Pectinate setæ rather long, with numerous pectinations or teeth, all on one plane. No aciculæ or spines to be seen. Uncini in general forcipate at the extremity. Tail furnished with two rather long cirri.

This is the general appearance presented by this species. In many points, however, there are variations from this normal structure. The setæ of the feet vary much in number, the greater number of both kinds (especially the pectinate setæ) being situated about the middle portion of the body. The two or three anterior pairs of feet, and the two last pair, have the uncini or hooklets changed into the appearance of the compound setæ of *Eunice*, the falcate appendage, however, being as it were soldered to the shaft and small—distinctly bidentate, as in *Eunice*. The uncini are generally two in number to each foot; but occasionally there are three, and generally one is smaller than the other. The caudal cirri show considerable variation also. In general there are two; but in one or two specimens examined there were three distinct cirri, and in one specimen one of the two cirri was divided, soon after it had sprung from the body, into two, or became, as it were, dichotomous.

From this variableness of the different portions of the body I have assigned to it its specific name.

Hab. St. Vincent's, West Indies, *L. Guilding.*

The genus *Hyalinœcia* of Malmgren was first established by Dr. Johnston, in his 'Catalogue of British Non-Parasitical Worms,' in 1865, under the name of *Northia*. Malmgren changes the name *Northia* to *Nothria*, and derives it from the Greek word νωθρὸς, *piger* (slow ?). He says that Dr. Johnston must have written it *Northia* in a mistake, unless he derived it from the word *North*, in the same way as Dr. Gray formed his genus *Fromia* (in Echinoderms) from the English preposition *from*. I suspect Dr. Johnston had no idea of deriving his genus *Northia* from the English word *North* (point of the compass), but that it was intended as a compliment to a person of the name of *North*.

In 1847 Dr. Gray named a genus of Mollusca *Northia*, taking as the type a species of Nassa (*N. Northia*), and so called it in honour of a person of the name of *North*. As this genus of Mollusks takes precedence by far in point of time of Johnston's genus of Annelides, I think it advisable, though for a very different reason from that given by Malmgren, to adopt this naturalist's correction, and for the future write the name *Nothria*. Johnston takes the species *Onuphis tubicola* as the type of his genus *Northia*, and gives as his chief reason for forming the genus (separating it from *Onuphis*) the fact that the two species referred to it are destitute of pectinate branchiæ, which exist in the species of the genus *Onuphis* as adopted by Audouin and M.-Edwards, Grube, &c. For the *Northia tubicola* of Johnston, Malmgren forms the new genus *Hyalinœcia*, while as the type of the genus *Nothria* he adopts the second species of Johnston's *Northia*, the *N. conchylega* (*Onuphis conchylega* of Sars). But I cannot see why M. Malmgren adopts Dr. Johnston's genus and at the same time refuses to accept the species *tubicola* as the type. For my own part, I should have preferred retaining the genus *Nothria* for the species *tubicola*, and should have wished M. Malmgren had constituted a new genus for *conchylega*. The only generic difference between the two species, as far as I can see, consists in the presence of the two postoccipital cirri in *conchylega*, and their absence in *tubicola*. Johnston does not seem to have seen these cirri in the specimens of *conchylega* which he examined; and Sars, who originally described the species, takes no notice of them, either in his description or his figures (see Sars, Beskriv. og Jaktt. p. 61, tab. 10. fig. 28). Our British specimens of the species are unfortunately imperfect, those from Berwick Bay (Dr. Johnston's own specimens) consisting of tubes only and one fragment of the animal; while the specimens we possess from the sea off the Shetland Islands, collected by Mr. Jeffreys, are equally fragmentary, seven or eight specimens existing of the inferior half of the animal only, not one having the head or anterior portion of the body entire.

It is just possible, therefore, that the *Northia conchylega* described and figured by Sars and Johnston may turn out to be a distinct species from that described by Œrsted, Grube, Malmgren, &c., which not only possesses the postoccipital cirri, but, according to Œrsted's figure and description of *Onuphis Eschrichti* (considered to be synonymous with *conchylega* by Malmgren), has also bipartite branchiæ.

On the Natural History and Hunting of the Beaver (*Castor canadensis*, Kuhl) on the Pacific Slope of the Rocky Mountains, by ASHDOWN H. GREEN, Esq. With Supplementary Notes by ROBERT BROWN, Esq., F.R.G.S. (Communicated by JAMES MURIE, M.D., F.L.S.)

[Read November 5, 1868.]

I HAVE have been for three years almost constantly engaged in trapping beavers, so that what remarks I may have to make on their habits and history, though somewhat at variance with the stereotyped notions prevalent in compilations, are yet the result of my own independent observations.

About January their tracks may be seen in the snow near the outlet of the lakes where young fir trees grow. At this time they prefer young fir trees as food to any other kind of tree, the reason, doubtless, being that at this period the sap has not risen in the willow or alder (*Alnus oregana*). It is not often that females are caught in the spring; and the males seem to travel about, as the runs are not used so regularly as they are when the beavers are living near.

Some of the beavers become torpid during January, especially those living near lakes, swamps, or large sheets of water which are frozen. They do not lay in a store of sticks for winter use as stated by Capt. Bonville (Washington Irving's 'Adventures of Capt. Bonville'), as one day's supply of sticks for a single beaver would fill a house—and if a stick were cut in the autumn, before the winter was over it would have lost its sap, and would not be eaten by the beaver. A beaver never eats the bark of a tree that is dead, though he may gnaw a hard piece of wood to keep his teeth down. A little grass is generally found in the houses, but is used as a bed and not for food.

If February is an open month, the beavers begin to come out of their retreats, and frequent any running water near them; but it is generally March before the bulk of them come out of winter-quarters. When they come out they are lean; but their furs are still good, and continue so till the middle of May—though if a trapper thought of revisiting the place, he would not trap after April, so as to allow them to breed quietly.

About the end of March the beaver begins to "call." Both males and females "call" and answer one another. Sometimes on one "calling," half-a-dozen will answer from different parts of the lake. I have known beavers to "call" as late as August. Males fight during the rutting-season most fiercely. Hardly a skin is without scars; and large pieces are often bitten out of their tails. The beaver holds like a bull-dog, but does not snap. It shakes its head so as to tear. When trapped, it will face a man, dodge a stick, and then seize it, taking chips out of it at every bite. It seems to attack from behind.

The period of gestation is known with little certainty, as they are never trapped in summer. The female brings forth some time about the end of June; and it is a year before a beaver is full-

Contributions towards a Monograph of the Species of *Annelides* belonging to the *Amphinomacea*, with a List of the known Species, and a Description of several new Species (belonging to the group) contained in the National Collection of the British Museum. To which is appended a short Account of two hitherto nondescript Annulose Animals of a larval character. By W. BAIRD, M.D., F.R.S., &c. &c.

(PLATES IV., V., VI.)

[Read April 2, 1868.]

IN the preceding volumes of the 'Journal of the Linnean Society.' Vol. VIII. pp. 172-202, and Vol. IX. pp. 31-38, I have communicated two papers to the Linnean Society, entitled "Contributions towards a Monograph of the *Aphroditacea*." In most of the systems of arrangement of the *Annelides*, the species of the group *Amphinomacea* succeed those of the *Aphroditacea*; and I now propose following up those papers by some contributions towards a further knowledge of the species of *Amphinomacea* also.

The few species known to Pallas and Gmelin were all arranged in the genera *Aphrodita* and *Terebella*. Bruguière first separated them from *Aphrodita*, and formed a distinct genus to receive them, to which he gave the name *Amphinome*. These worms differ much from the *Aphroditacea*, by the want of those organs called *elytra*, and by the presence of an uninterrupted series of branchiæ, which occur on almost all the segments of the body, and which do not alternate, as in these latter, with cirri. Many of them are very long and present a play of fine iridescent colours; most of them are natives of tropical seas. Since the genus *Amphinome* was formed by Bruguière, great additions have been made, several new genera and even distinct families have been formed; and as our knowledge of the various species which form this group increases, it will no doubt be found necessary to form several more.

Group *AMPHINOMACEA*.

Amphinomeaceæ, *Johnston*.
Amphinomea, *Kinberg, Carus*.

Family I. AMPHINOMIDÆ.

Amphinomea, *Blainville, Grube, Schmarda, Carus, Ehlers, Quatrefages*.
Amphinomæa, *Latreille*.

Amphinomæ, *Savigny, Lamarck.*
Amphinomiens, *Audouin & M.-Edwards.*
Amphinomacea, *Kinberg, Carus, Van der Hoeven.*
Amphinomidæ, *Gosse, Ann. & Mag. Nat. Hist.* 1853.

The animals belonging to this family possess a fleshy-looking caruncle or crest on the back of the buccal or cephalic segment, which is rounded. Branchiæ occur on almost all the segments of the body, are double, but do not alternate with cirri, as in the *Aphroditidæ.* The setiferous tubercles composing the feet are arranged in two rows, and are more or less widely apart. The eyes are four in number. The antennæ, as in the *Aphroditidæ,* have usually been described as five in number—one median, two internal, and two external. Following Kinberg's terminology, in accordance with what I have said in the case of the *Aphroditidæ,* the median single antenna will be designated as the *tentacle,* the internal pair as *antennæ,* and the external pair as *palpi.* Sometimes (*Euphrosyne*) the antennæ and palpi are wanting.

In the 'Öfversigt af Kongl. Vetenskaps-Akademiens Förhandlingar,' 1857, and afterwards in the 'Fregatten Eugenies Resa,' Kinberg (including the genus *Euphrosyne,* which he places in a family by itself) enumerates seven genera; to this number he adds, in the 'Öfv. Kong. Vetens.-Akad.' 1860, another, which, however, may be doubtful. Ehlers adopts the seven genera of Kinberg; but Quatrefages limits the number to four, though he describes one which does not enter into Kinberg's enumeration. Grube, in his 'Familien der Anneliden,' describes four, the same number as Quatrefages, but introduces one to which that author does not give a generic place. Audouin and M.-Edwards only admit three.

Genus I. AMPHINOME.

Aphrodita, sp., *Pallas.*
Terebella, sp., *Gmelin.*
Amphinome, *Bruguière,* 1789; *Cuvier, M.-Edwards, Règne Anim.* ed. *Crochard; Grube, Schmarda, Kinberg, Carus, Van der Hoeven, Quatrefages.*
Amphinoma, *Blainville,* 1828?; *Audouin & M.-Edwards, Littoral de la France.*
Pleione, *Savigny,* 1828?; *Cuvier, Lamarck, Stannius, Guérin.*

Body long, with the segments rectangular; cephalic lobe small, caruncle small, heart-shaped; antennæ and palpi rising from first segment of body; branchiæ commencing on 3rd or 4th segment of body; arborescent, branches filiform; *some of the setæ of the dorsal feet subulate, serrate, others linear, smooth ; setæ of the ventral feet hooked, thick, short, few in number.* Anus situated on the dorsal side of the lower extremity.

There are no species belonging to this genus found in Great Britain, though one, *Amphinome vagans*, has been described by Savigny as found by the late Dr. Leach on the coast of England. The locality of this species, however, was doubted by Savigny himself at a later period, and the specimen was suspected by him to have been brought to Dr. Leach from the Atlantic Ocean amongst some fuci. This has now, on the authority of Kinberg, been satisfactorily established, specimens having been brought by Dr. Schlör from the South Atlantic. Quatrefages, in his late work on the Annelides, enumerates twenty-seven species, including three which belong to the genus *Notopygos* of Grube, and which are distributed amongst five of the genera of Kinberg. To this list one or two new species have now to be added.

Sp. 1. AMPHINOME ROSTRATA. (Plate IV. figs. 1 *a*, *b*.)

Aphrodita rostrata, *Pallas, Miscell. Zoolog.* 100, tab. 8. f. 14-18, 1766.
Terebella rostrata, *Gmelin, Linn. Syst. Nat.* 3113.
Amphinome tetraedra, *Bruguière, Encyclop. Méthod. art.* Amphinome, Atlas, tab. 61. f. 8-12 (copied from *Pallas*); *Cuvier, Dict. Sc. Nat. art.* Amphinome.
Amphinoma tetraedra, *Blainville, Dict. Sc. Nat. art.* Vers, p. 450; *Audouin & M.-Edwards, Ann. Sc. Nat.* tom. xxviii. p. 197, *Hist. Nat. Littoral de la France*, ii. p. 123.
Pleione tetraedra, *Savigny, Syst. des Annélid.* 60; *Lamarck, An. s. Vert.* 1st edit. v. 330, 2nd edit. v. 572; *M.-Edwards, Cuv. Règn. Anim.* ed. Crochard, tab. 8 bis. fig. 1, 1a-1 c.
Amphinome rostrata, *Grube, Famil. der Annelid.* 40 and 122; *Van der Hoeven, Handbuch der Zoologie*, i. 231, 1850; *Carus, Handbuch der Zoologie*, ii. 435, 1863; *Quatrefages, Hist. Nat. des Annélides*, i. 393.

Hab. Indian Seas (*Madras, Mus. Brit.*); Australia (*Mus. Brit.*); Rio Janeiro, *Kinberg.*

I have had figured the setæ or bristles of the dorsal and ventral row of feet (*vide* Plate I. fig. 1). The setæ of the dorsal feet (fig. 1 *a*) are considerably longer than those of the ventral row, are very numerous, capillary, and terminate in a fine point. For some distance below this point they are serrated on the margins. The setæ of the ventral feet (fig. 1 *b*) are strong, curved at the apex, which is rather blunt, and below this are gradually enlarged. They are horny-looking in structure and colour, are simple or not toothed or serrated, are fewer in number and are much larger than those of the dorsal row.

Sp. 2. AMPHINOME VAGANS.
  Terebella vagans, *Leach, MS.* (fide *Savigny*).
  Pleione vagans, *Savigny, Syst. des Annélides*, p. 60.
  Amphinoma vagans, *Blainville, Dict. Sc. Nat. art.* Vers; *Audouin & M.-Edwards, Littoral de la France*, ii. 122.
  Amphinome vagans, *Grube, Famil. der Annelid.* 41 & 122; *Kinberg, Öfvers. Kong. Vetensk.-Akad. Förhand.* 1857, p. 12; *Fregatt. Eugen. Resa, Zoologi*, tab. xi. f. 6; *Quatrefages, Hist. Nat. des Annelés*, i. 403.
  Hab. South Atlantic Ocean, lat. 5° S., long. 50° W., *Kinberg.*

Sp. 3. ?AMPHINOME PALLASII.
  Pleione tetraedra, *M.-Edwards, Cuvier, Règn. Anim.* ed. Croch. tab. 8 bis. f. 1, 1 *a*.
  Amphinome tetraedra, *Valenciennes, MS. Coll. du Mus.* fide **Quatrefages**.
  Amphinome Pallasii. *Quatrefages, Hist. Nat. des Annelés*, i. 394.
  Hab. The Azores and West Indies, *Quatrefages.*

The chief differences between this species and *A. rostrata* appear to be the form of the branchiæ, which are divided into four or five separate branches, each rising from a particular root, and the shape and appearance of the caruncle.

Sp. 4. AMPHINOME LUZONLÆ.
  Amphinome Luzoniæ, *Kinberg, Öfvers. af Kong. Vetensk.-Akad. Förhandl.* 1857, p. 12; *Fregatt. Eugen. Resa, Zoologi, Annulat.* tab. xi. f. 7 *u*-7 *x*.
  Hab. West coast of Island of Luzon, *Werngren* fide *Kinberg.*

Sp. 5. AMPHINOME JUKESI, sp. nov. (Plate IV. figs. 2*a*, *b*.)
  Corpus utrinque attenuatum, quadratum, e segmentis 50 constans. Caruncula parva, cordiformis. Tentaculum breve, latum. Branchiæ parvæ, ramis subnumerosis. Setæ pedum dorsalium capillares, subulatæ, simplices. Setæ pedum ventralium breves, crassæ, simplices, apice incurvato, obtuso.
  Long. tres uncias æquans.
  Hab. Raine's Islet, North coast of Australia, *J. B. Jukes* (*Mus. Brit.*); ? China (in bad condition), *T. Lay, Esq.* (*Mus. Brit.*).

Worm about 3 inches in length, consisting of about 50 segments; of a quadrate shape, and narrower at each extremity. Caruncle small, heart-shaped. Tentacle short, flat and rather broad. Branchiæ small, of very short but rather numerous ramifications. Skin of the ventral surface of body coarsely wrinkled.

Feet prominent. Setæ of the upper or dorsal tuft (fig. 2 *a*) finely capillary, terminating in an acute long point. They are nearly quite simple, are indistinctly covered with minute prickles for a short distance below the apex, but have no serrations or teeth on their edge, and are five or six times longer than those of the lower or ventral tuft (fig. 2 *b*), which are short, stout, curved at the apex, which is rather blunt, but quite simple or free from serrations. In the shape of the caruncle and in the bristles of the lower or ventral tuft of the feet this species approaches near to *A. rostrata*, but it differs in the structure of the bristles of the upper or dorsal tuft. In this species they are capillary, finely acuminated, and nearly quite simple, whilst in *rostrata* they are stouter, and finely but distinctly serrated on both margins. The bristles of the lower or ventral tuft, again, in *jukesi*, are shorter, more curved, and not quite so horny in appearance as in *rostrata*. This species, too, is much smaller than *rostrata*, and of even a more quadrate or square shape.

Sp. 6. ?AMPHINOME CARNEA.

Amphinome carnea, *Grube et Œrsted, Annulat. Œrsted.* p. 26; *Quatrefages, Hist. Nat. des Annelés,* i .404 (quoted in synonyms, by mistake, *Amphinome rosea*).

Hab. Santa Cruz, *Œrsted.*

Genus II. HERMODICE.

Aphrodita, sp., *Pallas.*
Amphinome (part.), *Bruguière et auctorum.*
Pleione, sp., *Grube.*
Hermodice, *Kinberg, Öfvers. af Kong. Vetensk.-Akad. Förhand.* p. 11, 1857; *Fregatt. Eugen. Resa, Zoologi,* p. 32; *Ehlers, Die Borstenwürmer,* 64; *Carus, Handb. d. Zoologie,* ii. 435.

Body long, with the segments rectangular; cephalic lobe large. Caruncle large and lobed on each side. Branchiæ commence on the second segment. *Dorsal setæ, some subulate and serrate, others linear and smooth; ventral setæ serrated at the apex.*

Sp. 1. HERMODICE CARUNCULATA. (Plate IV. figs. 3 *a, b.*)

Millepeda marina Amboinensis, *Seba, Thes. rar. Nat.* tom. i. p. 131, tab. 81. no. 7, 1734-1765.

Nereis gigantea, *Linnæus, Syst. Nat.* ed. 12. tom. i. part 2. p. 1086. no. 10, 1766.

Aphrodita carunculata, *Pallas, Miscell. Zool.* pp. 102-106, tab. viii. f. 12-13, 1766.

Terebella carunculata, *Gmelin, Linn. Syst. Nat.* tom. i. part 6. Vermes, p. 3113. no. 5, 1789.
Amphinome carunculata, *Bruguière, Enc. Méth. art.* Amphinome, p. 46, *Atlas*, tab. 60. f. 6–7 (copied from *Pallas*), 1789; *Cuvier, Dict. des Sc. Nat. art.* Amphinome, tom. ii. p. 72; *Grube, Famil. der Annelid.* pp. 40 & 122; *Quatrefages, Hist. Nat. des Annelés,* i. 395.
Amphinoma carunculata, *Blainville, Dict. Sc. Nat. art.* Vers; *Audouin & M.-Edwards, Littoral de la France,* ii. 123.
Pleione carunculata, *Savigny, Syst. des Annélides,* p. 61; *Lamarck, An. s. Vert.* 1st edit. v. 330, 2nd edit. v. 572; *Cuvier, Règne Anim.* iii. 199, ed. *Crochard, Annélides,* tab. 8. f. 4, 4A; *Grube, De Pleione carunculatá; Treviranus, Beob. aus der Zoologie,* p. 53, tab. xi.
Hermodice carunculata, *Kinberg, Öfvers. Kongl. Vetensk.-Akad.* 1857, p. 13; *Carus, Handbuch der Zoologie,* ii. 435.

Hab. Seas of America, West Indies, St. Vincent's, West Indies, *Landsdown Guilding* (*Mus. Brit.*), West Indies, *Coll. Reid* (*Mus. Brit.*), St. Thomas's, West Indies (*Mus. Brit.*), Mediterranean, *Miller* (*Mus. Brit.*). The setæ of the dorsal row of feet (fig. 3*a*) are longer than those of the ventral feet, are very finely capillary, especially fine at the apex, and are all quite simple. The setæ of the ventral feet (fig. 3*b*) are numerous, fine, nearly capillary, but slightly curved at the apex, which is rather obtuse. For a short distance below the apex there are several very fine teeth or serræ, about 12 in number, on its inner margin; and a very short distance below these there is a prominent tooth on the inner edge.

Sp. 2. HERMODICE STRIATA.
Hermodice striata, *Kinberg, Öfvers. af Vetensk.-Akad. Förhandl.* 1857, p. 13; *Fregatt. Eugen. Resa,* tab. 12. f. 8, 8B–8G; *Carus, Handbuch der Zoologie,* ii. 435.

Hab. Eimeo, Pacific Ocean, among corals, near the shore, *Kinberg.*

Sp. 3. HERMODICE NIGROLINEATA, sp. nov.
Segmenta buccalia quinque. Branchiæ parvæ, sessiles, ramis paucis filiformibus. Caruncula magna, corrugata. Setæ pedum dorsalium omnes lineares, læves; setæ pedum ventralium bifidæ, ramo altero, brevissimo, dentem simulante, ramo altero longiore, apice breviuncinato, intus serrato. Dorsum corrugatum, segmenta singula, ad infimam partem, linea nigra notata. Tentaculum, antennis et palpis multo longior. Cirrus pedis dorsalis elongatus, gracilis.

Long. 2 unc. et 3 lin., lat. 3 lin.

Hab. Coast of Asia Minor, *R. M'Andrew.* On the submarine telegraph-cable, near Alexandria. Madeira, *Mr. Masson* (*Mus. Brit.*).

The tentacle is much longer than the antennæ or palpi, both of which latter are small. The caruncle resembles, in comparative size and in its corrugated character, that organ in *H. carunculata*, extending to the fourth segment of the body. The cirrus of the dorsal feet is longer than the setæ and rather slender. The setæ of the dorsal feet are all slender, linear, and quite smooth; those of the ventral feet are rather stouter; a short distance from the point, which is slightly curved or hooked, there is a tooth or short branch springing from it, and the space of the longer branch between this tooth and the point is rather strongly serrate on the inner side. The branchiæ are very small, and consist of only a a few filiform branches, from five to eight in number. The skin of the back is somewhat corrugated, and each segment has at its lower portion, near the junction of the following segment, a black line running across it, which is more strongly marked in the centre. Some of the specimens we possess were collected by Mr. M'Andrew on the coast of Asia Minor: one was taken from the submerged telegraph-cable, near Alexandria, when hauled up for examination; and others were collected by Mr. Masson in the sea of Madeira.

Sp. 4. HERMODICE DIDYMOBRANCHIATA.

Amphinome didymobranchiata, *Baird, Transact. Linn. Soc.* tom. xxiv. tab. 45. f. 1-7, 1864.

Hab. Island of Ascension, *Watson* (*Mus. Brit.*).

Sp. 5. HERMODICE SANGUINEA.

Amphinome sanguinea, *Schmarda, Neue wirbell. Thiere*, i. 2. pp. 140-141. fig. xylogr., tab. 34. f. 289; *Quatrefages, Hist. Nat. Annelés*, i. 405.

Hab. Jamaica, *Schmarda.*

Sp. 6. ? HERMODICE SAVIGNYI.

Amphinome Savignyi, *Brullé, Expd. de Morée, Zool.* tom. iii. p. 398. tab. 53. f. 1A-C; *Audouin & M.-Edwards, Littor. de la France*, ii. 124; *Quatrefages, Hist. Nat. Annelés*, i. 402.

Hab. Metana, coast of Sicily, *Brullé.*

Genus III. EURYTHOË.

Eurythoë, *Kinberg, Öfvers. af Kongl. Vetensk.-Akad. Förhandl.* p. 13, 1857; *Fregatt. Eugen. Resa, Zoologi,* p. 32; *Ehlers, Die Borstenwürm.* 64; *Carus, Handb. d. Zool.* ii. 435.

Pleione (part.), *Savigny*.
Amphinome (part.), *auctorum*.

Body long, with the segments rectangular; caruncle of middling size and minutely lobed. *Dorsal setæ, some linear, subarticulate, others subbifid, serrate, with one branch very short, rarely linear; ventral setæ bifid.*

Sp. 1. EURYTHOË ALCYONIA.

Pleione alcyonia, *Savigny, Syst. des Annélides*, p. 62; *Annélides gravés*, tab. 2. fig. 3; *Lamarck, An. s. Vert.* 1st ed. v. 331, 2nd ed. v. 572; *Blainville, Atlas Dict. Sc. Nat.* tab. vii. f. 2, 2A (copied from *Savigny*); *Guérin, Icon. Règne An. Annélides*, p. 4 (text); *Cuvier, Règne Anim.* iii. 199; *M.-Edwards, Cuv. Règn. Anim.* ed. Crochard, tab. 8 bis. f. 2 (copied from *Savigny*).

Pleyone alcyonia, *Guérin, Icon. R. An. Annélides*, tab. 4. f. 2, 2A–2c.

Amphinome alcyonia, *Blainville, Dict. Sc. Nat. art.* Vers; *Audouin & M.-Edwards, Littoral de la France*, ii. 124, tab. 22. f. 5 (copied from *Savig.*).

Amphinome alcyonia, *Grube, Famil. der Annelid.* pp. 40 & 122; *Quatrefages, Hist. Nat. des Annelés*, i. 401.

Hab. Red Sea, Dr. *Rüppell* (*Mus. Brit.*).

I refer this species to the genus *Eurythoë*. The caruncle is somewhat lobed at the edges. The setæ of the dorsal feet are, some linear, subarticulated in several places, others linear and subbifid, the terminal branch long and slender, whilst a third set are stouter and serrated. The setæ of the ventral feet are considerably stouter than the dorsal setæ, and are all bifid and quite smooth.

Sp. 2. EURYTHOË COMPLANATA. (Plate IV. figs. 4*a*, *b*.)

?Nereis tentaculis binis tripartitis, &c., *Brown, Hist. of Jamaica*, p. 395, tab. 39. f. 1 *.

Aphrodita complanata, *Pallas, Miscell. Zool.* 110, tab. 8. f. 19–26, 1766.

Terebella complanata, *Gmelin, Linn. Syst. Nat.* i. part 6. Vermes, p. 3113. no. 4.

Amphinome complanata, *Bruguière, Encycl. Méthod. art.* Amphinome, *Atlas*, tab. 60. f. 8–13 (copied from *Pallas*); *Grube, Famil. der Annelid.* pp. 40 & 122: *Quatrefages, Hist. Nat. des Annelés*, p. 403.

Amphinoma complanata, *Blainville, Dict. Sc. Nat. art.* Vers; *Audouin & M.-Edwards, Littoral de la France*, ii. 124.

Pleione complanata, *Savigny, Syst. des Annélides*, p. 62; *Cuvier, Règne Anim.* iii. 199; *Lamarck, An. s. Vert.* 1st ed. v. 331, 2nd edit. v. 573.

* Brown confounds this worm with the *Teredo* or shipworm! As Pallas conjectures, the specimen he had for inspection might probably have been taken burrowing in one of the holes made by the *Teredo*.

Hab. St. Vincent's, West Indies, *Guilding* (*Mus. Brit.*); St. Thomas's, West Indies, *Sallée* (*Mus. Brit.*, in bad condition); Eastern Seas (*Mus. Brit.*); Raine's Island, and Sir C. Hardy's Island, north coast of Australia, *Jukes* (*Mus. Brit.*); Zanzibar, *Dr. Kirk* (*Mus. Brit.*).

Notwithstanding the difference of habitat between the West Indies, north coast of Australia, and Zanzibar, I can see nothing to separate the two sets of specimens, except the greater size of the Australian. The specimens we possess from the Eastern seas are of about the same size as those from St. Vincent's, West Indies. The setæ of the dorsal and ventral feet are very nearly similar to those of the preceding species. Those of the dorsal row (fig. 4*a*) are numerous, capillary, but curiously and distinctly toothed or serrate on the edge. The apex is sharp-pointed, the teeth or serræ extend from it to some distance below it, are about 26 in number, and are harpoon-shaped. The setæ of the ventral feet (fig. 4*b*) are much fewer in number, and are stouter and shorter than those of the dorsal row. They are bifurcated near the apex and are quite simple or free from teeth or serrations.

Sp. 3. EURYTHOË HEDENBORGI.

Eurythoë Hedenborgi, *Kinberg, Öfvers. af Kongl. Vetensk.-Akad. Förhandl.* 1857, p. 13.

Hab. ——? From the collection of Dr. Hedenborg, *Kinberg.*

Sp. 4. EURYTHOË SYRIACA.

Eurythoë syriaca, *Kinberg, Öfvers. af Kongl. Vetensk.-Akad. Förhandl.* 1857, p. 13; *Carus, Handbuch der Zoologie.* ii. 435.

Hab. Coast of Syria, *Hedenborg* fide *Kinberg.*

Sp. 5. EURYTHOË CHILENSIS.

Eurythoë chilensis, *Kinberg, Öfvers. af Kong. Vetensk.-Akad. Förhandl.* 1857, p. 13; *Fregatt. Eugen. Resa, Zoologi. Annulat.* tab. xii. f. 9A–9x.

Hab. Near Valparaiso, depth of 7 fathoms, *Kinberg.*

Sp. 6. EURYTHOË CAPENSIS.

Eurythoë capensis, *Kinberg, Öfvers. af Kongl. Vetensk.-Akad. Förhandl.* 1857, p. 13; *Fregatt. Eugen. Resa, Zool. Annulat.* tab. xii. f. 10B, 10F, 10G.

Hab. Cape of Good Hope, *Wahlberg* fide *Kinberg.*

Sp. 7. EURYTHOË PACIFICA.

Eurythoë pacifica, *Kinberg, Öfvers. af Kongl. Vetensk.-Akad. Förhandl.*

1857. p. 14; *Fregatt. Eugen. Resa, Zoologi, Annulat.* tab. xii. f. 11A–11x; *Carus, Handb. der Zool.* ii. 435.

Hab. Pacific Ocean, near Eimeo and Foua Islands, amongst corals, *Kinberg.*

Sp. 8. EURYTHOË CORALLINA.

Eurythoë corallina, *Kinberg, Öfvers. Kongl. Vetensk.-Akad. Förhandl.* 1857, p. 14; *Fregatt. Eugen. Resa, Zoologi, Annulat.* tab. xii. f. 12B–12u.

Hab. Pacific Ocean, amongst corals on shores of islands Eimeo, Tahiti, and Oahu near Honolulu, *Kinberg.*

Sp. 9. EURYTHOË KAMEHAMEHA.

Eurythoë Kamehameha, *Kinberg, Öfvers. af Kongl. Vetensk.-Akad. Förhandl.* 1857, p. 14; *Fregatt. Eugen. Resa, Zool. Annulat.* tab. xii. f. 13c, f. g.

Hab. Harbour of Honolulu, amongst dead corals at 2 fathoms, *Kinberg.*

Sp. 10. EURYTHOË SMARAGDINA.

Amphinome smaragdina, *Schmarda, Neue wirbell. Thiere*, i. 2. p. 140. fig. xylogr., tab. 34. f. 288; *Quatrefages, Hist. Nat. des Annelés*, i. p. 405.

Hab. Jamaica, *Schmarda.*

Sp. 11. ?EURYTHOË LATISSIMA.

Amphinome latissima, *Schmarda, Neue wirbell. Thiere*, i. 2. p. 141. fig. xylogr., tab. 34. f. 291, 291A; *Quatrefages, Hist. Nat. Annelés*, i. 405.

Hab. Ceylon, *Schmarda.*

Sp. 12. EURYTHOË LONGICIRRA.

Amphinoma longicirra, *Schmarda, Neue wirbell. Thiere*, i. 2. p. 142. fig. xylogr., tab. 34. f. 292; *Quatrefages, Hist. Nat, Annelés*, p. 405.

Hab. Ceylon, *Schmarda.*

Sp. 13. EURYTHOË INDICA.

Amphinome indica, *Schmarda, Neue wirbell. Thiere*, i. 2. p. 142. fig. xylogr., tab. 35. f. 294; *Quatrefages, Hist. Nat. Annelés*, i. 405.

Hab. Ceylon, *Schmarda.*

Sp. 14. EURYTHOË JAMAICENSIS.

Amphinome (Notopygos?) Jamaicensis, *Schmarda, Neue wirbell. Thiere*, i. 2. p. 143. fig. xylogr.; *Quatrefages, Hist. Nat. Annelés*, i. 406.

Hab. Jamaica, *Schmarda.*

Sp. 15. EURYTHOË ENCOPOCHÆTA.

Amphinome encopochæta, *Schmarda, Neue wirbell. Thiere*, i. 2. p. 143. fig. xylogr., tab. 35. f. 293; *Quatrefages, Hist. Nat. Annelés*, i. 406.

Hab. Ceylon, *Schmarda.*

Sp. 16. EURYTHOË MACROTRICHA.
Amphinome macrotricha, *Schmarda, Neue wirbell. Thiere,* i. 2. p. 144. fig. xylogr., tab. 34. f. 290; *Quatrefages, Hist. Nat. Annelés,* i. 406.

Hab. Jamaica, *Schmarda.*

Sp. 17. ? EURYTHOË CLAVATA. (Plate IV. figs. 5 *a*, *b*.)

Corpus depressum, subquadratum, ad extremitates utrinque attenuatum, e segmentis 55 s. 56 constans. Caruncula ovalis, mediocris, segmenta tria prima tegens. Setæ pedum dorsalium ad apicem curvatæ, dilatatæ seu clavatæ, simplices; setæ pedum ventralium bifurcæ, læves. Branchiæ parvæ, ramis ramulisque numerosis.

Long. tres uncias æquans.

Hab. ——? (*Mus. Brit.*).

Worm about 3 inches in length, of a flatly subquadrate shape, and consisting of about 55 or 56 segments. Body of a very dark colour; bristles of feet, especially those of ventral or lower tuft, light or yellowish coloured, and tipped at the points with dark brown. The caruncle is rather large, of an oval shape, the edges not rolled up, and extending over the first three segments.

The specimen under observation is thickest and broadest in the middle; narrowed at each extremity, but becoming suddenly contracted about the thirty-fourth segment, and appearing much rower at the posterior than at the anterior extremity.

The setæ of the upper or dorsal tuft (fig. 5 *a*) are slightly curved and obtuse or rather club-shaped at the extremity, which is simple and not toothed. Those of the lower or ventral tuft (fig. 5 *b*) are bifurcate, the rami not serrated, and the points rather blunt.

The branchiæ are small, numerously ramified, and of a dark colour.

Only one specimen of this species has occurred, and we have no history as to its habitat.

Genus IV. LYCARETUS.

Lycaretus, *Kinberg, Öfvers. af Kongl. Vetensk.-Akad. Förhandl.* 1867, p. 53.

Body long, depressed, segments rectangular; cephalic lobe rounded, caruncle elongate, rather smooth; eyes, tentacle, antennæ, and palpi as in *Amphinome*. Branchiæ commencing from the third segment. Dorsal cirri on each side single. *Setæ of dorsal feet capillary, somewhat geniculate, some of them serrated; setæ of ventral feet bifid, the points unequal, smooth.*

Sp. 1. LYCARETUS NEOCEPHALICUS.
Lycaretus neocephalicus, *Kinberg, l. c.* p. 56.
Hab. West Indies, Bartholomew Island, *Lovén* &c.

Genus V. LIRIONE.

Lirione, *Kinberg, Öfvers. af Kongl. Vetensk.-Akad. Förhandl.* p. 12, 1857; *Fregatt. Eugen. Resa, Zoologi,* p. 32; *Ehlers, Die Borstenwürm.* p. 64; *Carus, Handbuch der Zoologie,* ii. 435.
Amphinome, pars, *auctorum*.

Body elongate, with the segments oval and large; cephalic lobe rounded, elevated. Antennæ rising from the cephalic lobe, palpi from the first segment of body. Caruncle elongate. Branchiæ placed near the apex of the dorsal feet. Dorsal cirri, two on each side. *Setæ all alike, bifid, smooth.*

Sp. 1. LIRIONE SPLENDENS.
Lirione splendens, *Kinberg, Öfvers. af Kongl. Vetensk.-Akad. Förhandl.* 1857, p. 12; *Fregatt. Eugen. Resa, Zool. Annulat.* tab. xi. f. 4A–4X; *Carus, Handbuch der Zoologie,* ii. 435.
Hab. Near the Island of Tahiti amongst corals, at a foot depth, *Kinberg*.

Sp. 2. LIRIONE MACULATA.
Lirione maculata, *Kinberg, Öfvers. af Kongl. Vetensk.-Akad. Förhandl.* p. 12; *Fregatt. Eugen. Resa, Zool. Annulat.* tab. xi. f. 5B–5X.
Hab. Coast of islands near Panama, *Kinberg*.

Sp. 3. LIRIONE RAYNERI, sp. nov. (Plate IV. figs. 6, *a, b.*)
Corpus elongato-fusiforme, utrinque attenuatum, e segmentis triginta duobus seu triginta tribus constans. Caruncula magna, sextum segmentum attingens, crista media alta laminaque laterali majuscula (transverse valde plicatis) ornata. Cirri dorsales bini. Tuberculi setiferi dorsales basi linea nigra circumdati. Branchiæ a segmento quinto orientes, breves, ramis filiformibus, basi nigris, divisæ. Setæ dorsales et ventrales conformes, apice bifido, simplices; setæ ventrales breviores. Anus dorsalis, segmento vicesimo secundo situs. Appendices anales breves, obtusæ, binæ.

Long. uncias duas et quartam partem æquans.

Hab. Reefs off the north-east coast of Australia, *Rayner* (*Mus. Brit.*).

Body somewhat elongately fusiform, narrower at each extremity, about 2¼ inches long, and composed of 32 or 33 segments. Beneath or on the ventral surface it is of a light brown colour, but the back is violet and marked with a number of white lines crossing each other in various directions. The caruncle is large, extending to the sixth segment of the body. It is apparently composed of three portions, which are almost separate from each other. The centre portion, or crest, is detached from the lateral portions throughout its whole length, except at the two extremities. All three portions are strongly wrinkled. The setiferous tubercles are prominent, the dorsal being encircled at the base with a black ring. The branchiæ arise from the fifth segment, are placed upon the base of the dorsal setiferous tubercles, and are composed of a tuft of short cirriform branchlets or filaments, about from 20 to 55 in number. The anus is placed on the back, on the twenty-second segment, and in the centre of a rounded fleshy caruncle. The setæ or bristles of both dorsal and ventral tufts (fig. 5, *a*, *b*) are long, capillary, and sharply bidentate a little way below the apex, which is simple, the tooth being sharp and erect. The ventral setæ (fig. 5, *b*) are rather shorter than those of the dorsal tuft. The dorsal cirri are double,—one, the most dorsal, is short and subulate, about the length of the branchial filaments; the other, the most ventral in position, is much longer, and composed of two joints, the basal much the stouter of the two. The ventral cirri are single, and about the same length as the most dorsal of the dorsal cirri.

Only one specimen was brought to the Museum. It was taken by F. Rayner, Esq., Surgeon of H.M.S. 'Herald,' to whom I have dedicated this fine species.

Genus VI. LINOPHERUS.

Linopherus, *Quatrefages.*
Amphinome, sp. *Peters, Grube.*

Body linear. Head as in *Amphinome*. Caruncle very small. Feet in two rows, apart. Branchiæ cirriform, the cirri of which they are composed being either simple or bifurcate.

Sp. 1. LINOPHERUS INCARUNCULATUS.

Amphinome incarunculata, *Peters*; *Grube, Beschr. neuer od. wenig bekannt. Annelid. in Troschel, Archiv der Naturg.* 1860, p. 77.

Linopherus incarunculata, *Quatrefages, Hist. Nat. Ann.* i. 407.
Hab. West Africa, *Peters, Grube.*

Genus VII. NOTOPYGOS.

Notopygos, *Grube, Famil. der Annelid.* 121, 1851; *Beschr. neu. od. wenig bekannt. Annelid., in Archiv der Naturg.* 1855, p. 93; *Annulat. Œrsted.* p. 27; *Ehlers, Die Borstenwürmer,* p. 64.
Notopygus, *Kinberg, Öfvers. af Kongl. Vetensk.-Akad. Förhandl.* 1857, p. 11; *Fregatt. Eugen. Resa, Zoologi,* p. 32 (char. emend.); *Carus, Handb. d. Zool.* ii. 435.
Amphinome, sp., *Quatrefages.*

"Body of an oval shape, with large oval segments; cephalic lobe depressed; antennæ and palpi rising from first segment of body; caruncle elongate; branchiæ cirrated, placed at the apex of the dorsal feet; cirrus of dorsal feet single, *setæ of dorsal feet bifid, the longer branch lightly serrated inwardly,* anal appendages double." Char. emend., *Kinberg.*

Sp. 1. NOTOPYGOS CRINITUS.
Notopygos crinita, *Grube, Famil. der Annelid.* p. 40; *Neuer od. wen. bekannt. Annelid. Troschel, Archiv,* tom. xli. 1855, p. 93.
Notopygus crinitus, *Kinberg, Öfvers. af Kongl. Vet.-Akad. Förhand.* 1857, p. 11; *Fregatt. Eugen. Resa, Zoolog. Annulat.* tab. xi. f. 3A–3x; *Carus, Handb. der Zoologie,* ii. 435.
Amphinome crinita, *Quatrefages, Hist. Nat. Annelés,* tom. i. p. 403.
Hab. Near Island of St. Helena, in 80 fathoms, *Kinberg.*

Sp. 2. NOTOPYGOS ORNATUS.
Notopygos ornata, *Grube, Annulat. Œrsted.* p. 27.
Amphinome ornata, *Quatrefages, Hist. Nat. Annelés,* tom. i. p. 404.
Hab. Puntarenas, in Costa Rica, *Grube.*

Species of Amphinomidæ *which cannot as yet be referred to their proper genera.*

Sp. 1. AMPHINOME ÆOLIDES.
Pleione æolides, *Savigny, Syst. des Annélides,* p. 62; *Lamarck, An. s. Vert.* 1st edit. v. 330, 2nd edit. v. 572.
Amphinome æolides, *Blainville, Dict. Sc. Nat. art.* Vers; *Audouin & M.-Edwards, Littoral de la France,* i. 124.
Amphinome æolides, *Grube, Famil. der Annelid.* pp. 40 & 122; *Quatrefages, Hist. Nat. Annelés,* i. 397.
Hab. West Indies, *Quatrefages.*

Sp. 2. AMPHINOME ABHORTONI.
Amphinome abhortoni, *Valenciennes, MS.*? ; *Quatrefages, Hist. Nat. Annelés*, i. 397.
Hab. Isle of France, *Quatrefages.*

Sp. 3. AMPHINOME BRUGUIERESI.
Amphinome Bruguieresi, *Quatrefages, Hist. Nat. Annelés*, i. 398.
Hab. Seychelles, *Quatrefages.*

Sp. 4. AMPHINOME FORMOSA.
Amphinome formosa, *Valenciennes, MS.*; *Quatrefages, Hist. Nat. Annelés*, i. 399.
Hab. Sandwich Islands, *Quatrefages.*

Sp. 5. AMPHINOME DENUDATA.
Amphinome denudata, *Quatrefages, Hist. Nat. Annelés*, i. 400.
Hab. New Caledonia, *Quatrefages.*

Sp. 6. AMPHINOME GAUDICHAUDI.
Amphinome Gaudichaudi, *Valenciennes, MS.*; *Quatrefages, Hist. Nat. Annelés*, i. 400.
Hab. Paëta, *Quatrefages.*

Sp. 7. AMPHINOME PALLIDA.
Amphinome pallida, *Quatrefages, Hist. Nat. Annelés*, i. 401.
Hab. —— ?

Sp. 8. AMPHINOME PAUPERA.
Amphinome paupera, *Grube & Œrsted, Annulata Œrstediana*, p. 26; *Quatrefages, Hist. Nat. Annelés*, i. 404.
Hab. Valparaiso, *Œrsted.*

Sp. 9. AMPHINOME STILIFERA.
Amphinome stilifera, *Grube, Besch. neuer oder wenig bekannt. Ann.* p. 78; *Quatrefages, Hist. Nat. Annelés*, i. 406.
Hab. —— ?

Sp. 10. AMPHINOME PELAGICA.
Amphinome pelagica, *Quoy & Gaimard, MS. in Mus. Paris*; *Audouin & M.-Edwards, Littoral de la France*, note at p. 124 ; *Grube, Famil. der Annelid.* p. 41.
Hab. Amboina, *Quoy & Gaimard.*

Genus VIII. CHLOIEA.

Aphrodita, sp., *Pallas.*
Terebella, sp., *Gmelin.*
Amphinome sp., *Bruguière, Cuvier.*

Chloeia, *Savigny, Cuvier, Blainville, Lamarck, Audouin & M.-Edwards, Risso, Grube, Carus, Van der Hoeven, Ehlers, Schmarda, Quatrefages, Kinberg.*

Body oval in shape, with the segments oval; antennæ and palpi rising from the first segment; caruncle elongate; *branchiæ bipinnate, placed at some distance from the apex of the feet*; cirrus of dorsal foot single; *setæ of dorsal feet serrate*; *setæ of ventral feet bifid*; anal appendages double. Eyes, as in *Amphinome*, 4 *.

Sp. 1. CHLOEIA FLAVA.

Aphrodita flava, *Pallas, Miscell. Zoolog.* 97, tab. viii. f. 7–11.
Terebella flava, *Gmelin, Linn. Syst. Nat.* i. part 6. p. 3114; ?*Krusenstern, Atlas*, tab. 88. f. 14–16.
Amphinome capillata, *Bruguière, Encyc. Méthod. art.* Amphinome, *Atlas*, tab. lx. f. 1–5 (copied from *Pallas*); *Cuvier, Règn. Anim.* iii. 198.
Chloeia capillata, *Savigny, Syst. des Annélides*, p. 58; *Lamarck An. s. Vert.* 1st edit. v. 329, 2nd edit. v. 570; *Audouin & M.-Edwards, Littoral de la France*, ii. 120, tab. ii B. f. 11–12; *M.-Edwards, Cuv. R. An.* ed. Croch. tab. ix. f. 1; *Van der Hoeven, Handbuch der Zoologie*, i. 231; *Carus, Handbuch der Zoologie*, ii. 435.
Amphinome flava, *Cuvier, Dict. Sc. Nat. art.* Vers; *Blainville, Atlas, Dict. Sc. Nat.* tab. vii. figs. 1–1A–1c.
Chloeia flava, *Blainville, Dict. Sc. Nat. art.* Vers; *Grube, Famil. der Annelid.* p. 40; ?*Quatrefages, Hist. Nat. des Annelés*, i. 386; *Kinberg, Fregatt. Eugen. Resa, Zoolog. Annulat.* tab. xi. f. 1B–1x.
Chloeia incerta, *Quatrefages, Hist. Nat. Annelés*, i. 398. no. 2.

Hab. Chinese and Indian Seas (*Mus. Brit.*); Australia (*Mus. Brit.*).

The *Chloeia incerta* of Quatrefages belongs undoubtedly to this species, the only reason for asserting the *Chloeia flava* of Pallas is not the *Chloeia capillata* of M.-Edwards being the difference of the setæ of the feet. We possess, in the collection of the British Museum, nine specimens of what appears to me to be the true *Chloeia flava* of Pallas, the setæ of the feet of which all

---

* Savigny, who established the genus *Chloeia*, distinctly asserts that the species have only *two* eyes. In this he has been followed, apparently without examination, by Lamarck, Audouin and M.-Edwards, Risso, Grube, Van der Hoeven, Schmarda, and Quatrefages. Pallas, Gmelin, Cuvier, and Carus do not notice the number, but Kinberg has recognized the fact that they are endowed with *four*. This I have also ascertained to be the case in *Chloria flava, tumida, pulchella, parva*, and *spectabilis*. Kinberg has distinctly figured *four* eyes in *C. flava* and *candida*: and I have little doubt four is the normal number of eyes in this genus.

agree with those of *Chloeia capillata* figured by M.-Edwards in figs. 1D, E of pl. 9, in the Crochard edition of the 'Règne Animal,' and by Kinberg in the 'Fregatt. Eugen. Resa,' plate xi. fig. 1, H, G. The description of these setæ by Savigny, " celles des rames supérieures simplement aiguës, les autres terminées par un pointe distincte," though agreeing better with Quatrefages's description of these organs, is not sufficiently precise to determine the question of the distinction of the two species. If the species quoted as *Chloeia flava* by Quatrefages be distinct, I think it ought to be formed into a different species and deserves better the name of *incerta*.

Sp. 2. ? CHLOEIA QUATREFAGESII.

Chloeia flava, *Quatrefages* (without any of the synonyms quoted by him), *Hist. Nat. des Annelés*, i. 386. no. 1.

Hab. Seas of China, *Quatrefages*.

Sp. 3. CHLOEIA CANDIDA.

Chloeia candida, *Kinberg, Öfvers. Kongl. Vetensk.-Akad. Förhandl.* 1857, p. 11; *Fregatt. Eugen. Resa, Zool. Annulat.* tab. xi. f. 2, 2A-2x; *Carus, Handbuch der Zoologie*, ii. 435.

Hab. A small species coming from the Island of St. Thomas (?West Indies), *Werngren* fide *Kinberg*.

Sp. 4. CHLOEIA FURCIGERA.

Chloeia furcigera, *Quatrefages, Hist. Nat. des Annelés*, i. 309.

Hab. Mauritius, *Quatrefages*.

The specimen from which Quatrefages drew up the description of this species is in such bad condition, he says, that he can only draw attention to the setæ of the feet, which are bifid in both dorsal and ventral feet.

Sp. 5. ? CHLOEIA INERMIS.

Chloeia egena?, *Grube, Beschr. neuer od. wenig bekannt. Ann.* 1855, p. 91.

Chloeia inermis, *Quatrefages, Hist. Nat. Annelés*, tom. i. p. 389.

Hab. New Zealand, *Quatrefages*.

Sp. 6. CHLOEIA EGENA.

? Chloeia egena, *Grube, Beschr. neu. od. wen. bekannt. Ann. in Archiv d. Naturg.* 1855, p. 91; *Quatrefages, Hist. Nat. Annelés*, tom. i. p. 391.

Hab. ——? (in the *Museum of St. Petersburg*), *Kinberg*.

This species is doubtful, according to Quatrefages, and, he says, may prove to be identical with his *C. inermis* (*vide* Sp. 13 of this list).

Sp. 7. CHLOEIA FUCATA.
Chloeia fucata, *Quatrefages, Hist. Nat. Annelés*, tom. i. p. 390.
Hab. Mascate, *Quatrefages*.

Sp. 8. CHLOEIA NUDA.
Chloeia nuda, *Quatrefages, Hist. Nat. Annelés*, tom. i. p. 390.
Hab. Amboina, *Quatrefages*.

Sp. 9. CHLOEIA VENUSTA.
Chloeia venusta, *Quatrefages, Hist. Nat. Annelés*, tom. i. p. 391.
Hab. Palermo, *Quatrefages*.

Sp. 10. CHLOEIA VIRIDIS.
Chloeia viridis, *Schmarda, Neue wirbell. Thiere*, tom. ii. p. 144. fig. xylogr., A–X. tab. 35. f. 295–305; *Quatrefages, Hist. Nat. Annelés*, tom. i. p. 392.
Hab. Coast of Jamaica, *Schmarda*.

Sp. 11. CHLOEIA TUMIDA, sp. nov. (Plate IV. figs. 7 *a–d*.)
Corpus tumidum, album, e segmentis triginta et sex constans. Oculi parvi. Caruncula mediocris, plicata. Branchiæ bipinnatæ, ab segmento quarto orientes. Oculi parvi. Cirri dorsales longiores quam ventrales, et tenuiores. Cirri anales crassi, breves. Setæ pedum dorsalium breviores quam ventralium, et parum numerosæ, paululum infra apicem tumidæ, extus prope apicem unidentatæ, intus valde serratæ (interdum simplices). Setæ pedum ventralium longæ capillares, bifidæ, ramo interno brevissimo, dentem, simulante, ad apicem simplices.

Long. uncias sex et dimidiam æquans; lat. (setis inclusis) uncias duas æquans.

Hab. India, *Leadbeater* fide *Leach* (*Mus. Brit.*).

The body of this worm is very tumid or swollen, quite white in all its parts and destitute of any markings or colour. It is much longer than broad, in length being 6½ inches, in breadth (including the setæ) 2 inches, and is composed of 36 segments. The skin or dorsal surface is wrinkled. The caruncle is moderately large and plicate, and the branchiæ are bipinnate, and commence from the fourth segment of the body. The eyes are very small. The dorsal cirri are longer and more slender than the ventral; the anal cirri are very stout, rather short, and of about the same thickness throughout their length. The bristles of the ventral or lower row of feet (fig. 7 *d*) are long, white, capillary, terminating in a rather sharp point, bifid, the inner ramus very short, more resembling a tooth, springing from a little below the apex, and are simple, not serrated. The bristles of the dorsal or upper row of feet (fig. 7 *a*, *b*, *c*) are fewer in number than those of the ventral row; they are curiously swollen out a little below the apical part, which terminates in a sharp point with a small tooth on its outer edge a very short distance beneath the point, and, with the exception of those of the second or third feet (fig. 7 *b*), they have a row of serrations or teeth on the inner margins, the teeth being about 16 in number, slightly curved downwards or harpoon-shaped.

We possess only one specimen of this species, which formed part of the collection of the late Dr. Leach, and, on the authority of Mr. Leadbeater, who supplied him with the specimens, is said to come from India.

Sp. 12. CHLOEIA PARVA, sp. nov. (Plate IV. figs. 8 *a*, *b*.)

Corpus breve, angustum, utrinque attenuatum, e segmentis viginta et sex constans, atratum, segmentis omnibus medio dorsi linea nigra longitudinali et maculis atris transversis, signatis. Branchiæ parvæ, nigræ. Caruncula elongata, plicata, crista media linea nigra notata. Cirri dorsales nigri. Cirri ventrales pallidi. Setæ pedum dorsalium capillares, acute punctatæ, margine interno serratæ, serræ deorsum spectantes. Setæ pedum ventralium capillares, bifidæ, ramus internus brevissimus, dentem simulans.

Long. unciam unam æquans.

Hab. ———? (*Mus. Brit.*).

This worm is small, only about 1 inch in length, and narrow, attenuated at each extremity, especially posteriorly, the posterior extremity being much narrower than the anterior, and produced into a sharpish point. The body is composed of about 26 segments, is of a dark colour, and peculiarly marked on the back with black spots. Along the centre of the back, on each segment, there is a dark mark in the shape somewhat of the Roman **T**, or rather the Greek ϒ (upsilon). On each side there is also a dark mark which runs across each of the segments, and another encircling the ventral setiferous tubercle or foot. The branchiæ are small, simply branched, and are of a dark colour. The caruncle is elongate and much wrinkled or pleated, and its crest is surmounted with a black, waved line. The setæ of the dorsal tuft (fig. 8 *a*) are capillary, sharp-pointed, and serrated or toothed a little below the apex with harpoon-shaped teeth. The setæ of the ventral tuft (fig. 8 *b*) are very slender, capillary, and shortly bifurcate near the apex, the short branch more resembling a tooth. The dorsal cirri are of a black colour; the ventral are pale. The anal cirri are of a light colour, short, stout, and cylindrical. The habitat of this well-marked species is unknown. The animal, at first sight, resembles a species of *Hipponoa*.

Sp. 13. CHLOEIA SPECTABILIS, sp. nov.

Corpus rotundato-fusiforme, utrinque attenuatum, e segmentis triginta et quatuor constans, pallidum, albo punctatum. Caruncula elongata, angusta, quintum segmentum attingens. Cirri dorsales subulati, purpurei. Cirri ventrales albi. Setæ pedum dorsalium et ventralium capillares, lineares, simplices.

Long. uncias duas et dimidiam æquans.

Hab. New Zealand, *Capt. Stokes* (*Mus. Brit.*).

Worm about 2½ inches long. Body rounded-fusiform in shape, thicker in the centre and narrower at each extremity, but the posterior extremity narrower than the anterior; composed of about 34 segments. It is of a light colour, and the whole body above and below is dotted all over with numerous small, white, round spots varying in size. The caruncle is long and narrow, extending over 4 or 5 segments. Ventral cirri white; dorsal cirri long and subulate and of a beautiful purple colour. Branchiæ simply branched. Bristles of both ventral and dorsal feet capillary, slender and simple, those of the dorsal tuft longer and stouter than the ventral.

This species, in its habitat and structure of the setæ of the feet, approaches the two species "*inermis*" of Quatrefages, and "*egena*" of Grube, but differs from both in minor details.

Sp. 14. CHLOEIA PUCHELLLA, sp. nov.

Corpus depressiusculum, fusiforme, e segmentis triginta duobus seu triginta et quinque constans. Dorsum rugosum, in medio segmentorum macula angusta nigra notatum. Oculi parvi. Caruncula elongata, crista linea nigra insignis. Branchiæ pinnatifidæ, ramusculis atratis. Cirri dorsales elongati, subulati, atrati: cirri ventrales albi; cirri anales breves, cylindrici. Setæ pedum dorsalium infra apicem serratæ; setæ ventrales bifidæ, simplices.

Long. uncias duas æquans; lat. (setis inclusis) unciam dimidiam æquans.

Hab. Reefs off the north-east coast of Australia, *F. M. Rayner* (*Mus. Brit.*).

The body of this worm is rather narrow, of a depressed fusiform shape, about 2 inches long, and half an inch broad, including the setæ of feet. Segments of body from 32 to 35. Eyes small, the anterior pair larger than the posterior. Caruncle long, extending over a considerable number of segments, rugose, the summit of the crest marked with a waved dark line. Branchiæ finely and delicately pinnatifid, the main branch of a light colour, the branchlets very dark. Setæ of feet yellow; those of the ventral feet forming a tuft considerably larger than that of the dorsal, and in both formed like that of those of *Chloeia flava*—those of the dorsal tuft having their apex for a little way down on the inner side serrated with harpoon-shaped teeth, while those of the ventral tuft are furcate near the apex and simple. The back or dorsal surface of the body is wrinkled, of a pale delicate flesh-colour, and is marked on the middle of each segment, between the branchiæ, with a distinct rather broad line of a very dark colour—not square-shaped, as in *C. flava*, but extending in length to more than half the breadth of the segment. The dorsal row of setiferous tubercles is also marked with a dark line running partly across them just above the tuft of setæ. The ventral cirri are nearly white and finely pointed; the dorsal cirri are of a very dark colour, and are longer and more slender than the ventral. The anal cirri are short and cylindrical, and rather stout.

Var. *a. pallida.*

We are indebted to Mr. Jukes for a specimen of a worm which I consider to be only a variety of *Chloeia pulchella*. It is nearly colourless in all parts except the peculiar marks on the dorsal surface of the segments. The setæ of the feet are of a lighter yellow colour, but their structure is the same as those of *pulchella*. The worm is a little larger, and the colourless branchiæ are rather larger also than those of the type specimens.

Hab. Raine's Islet, North Australia, *J. B. Jukes* (*Mus. Brit.*).

Uncertain Species.

?Chlocia rupestris, *Risso, Hist. nat. Eur. mérid.* iv. 425.

This species evidently belongs to another family altogether. Quatrefages says it most probably belongs to the genus *Eunice*. M.-Edwards says he cannot refer it to any genus of the family Amphinomidæ, as Risso describes it as possessing *jaws*, a character which does not belong to any species of the family.

Genus IX. EUPHROSYNE.

Euphrosyne, *Savigny, Syst. des Annélides*, p. 63, 1809?; *Audouin & M.-Edwards, Grube, Johnston, Kinberg, Sars, Carus, Van der Hoeven, Schmarda, Quatrefages, Ehlers.*
Euphrosine, *Cuvier, R. An.; Lamk. An. s. Vert.*; *M.-Edwards, Cuv. R. A. ed. Croch.; Blainville.*
Euphrosina and Euphrosyna, *Audouin & M.-Edwards, Littoral de la France; Œrsted.*

Antennæ and palpi wanting. Cephalic lobe compressed. Caruncle elongate. Eyes two. Branchiæ ramose, several in each segment. Body oval, with the segments rounded. Anus placed on the dorsal aspect, with a longitudinal orifice and two appendages. Feet crest-shaped, placed transversely. *All the setæ bifid, those of dorsal tuft smooth, those of ventral tuft with the inner side of the rami serrated.*

A. *With both cirri and branchiæ on the sides of the segments.*

I. *Branchiæ all ramose.*

Sp. 1. EUPHROSYNE LAUREATA.
Euphrosyne laureata, *Savigny, Syst. des Annélides*, p. 63, *Annélides gravés*, tab. 2. fig. 1; *Grube, Fam. der Anneliden*, pp. 41 & 122; *Quatrefages, Hist. Nat. Annelés*, i. 409; *Ehlers, Die Borstenwürmer*, p. 65; *Audouin & M.-Edwards, Littoral de la France*, tom. ii. p. 127.
Euphrosine laureata, *Cuvier, R. An.* tom. iii. p. 199; *Blainville, Dict. Sc. Nat. art.* Vers, p. 453, Atlas, tab. 8. f. 1, 1A–1c (copied from Savigny); *Guérin, Icon. R. An.* tab. iv. bis. f. 1; *Van der Hoeven, Handbuch der Zoologie*, i. 231; *Carus, Handbuch der Zoologie*, ii. 435; *Lamarck, An. s. Vert.* 1st edit. v. 332, 2nd edit. v. 574; *M.-Edwards, Cuv. R. An. ed. Croch.* (texte), tab. 8. f. 3–3A (copied from Savigny); *Quatrefages, Hist. Nat. des Annelés*, tom. i. p. 409 (in list of synonyms).
Hab. Red Sea, *Savigny.*

Sp. 2. EUPHROSYNE MYRTOSA.
Euphosyne myrtosa, *Savigny, Syst. des Annélides*, p. 64, *Annélides gravés*, tab. ii. f. 2; *Grube, Famil. der Anneliden*, pp. 41 & 122; *Ehlers, Die Borstenwürmer*, p. 66; *Quatrefages, Hist. Nat. Annelés*, i. 409.
Euphrosyna myrtosa, *Audouin & M.-Edwards, Ann. Sc. Nat.* tom. xx. tab. iii. f. 6–8; *Littoral de la France*, ii. 128.
Euphrosine myrtosa, *Cuvier, R. An.* iii. 199; *Blainville, Dict. Sc. Nat. art.* Vers, p. 453; *Cuvier, Iconograph. R. Anim.* tab. iv. bis. f. 2;

Lamarck, An. s. Vert. 1st edit. v. 332, 2nd edit. v. 574; *Quatrefages, Hist. Nat. Annelés,* i. 409.

Hab. Gulf of Suez, Red Sea, *Savigny.*

Sp. 3. EUPHROSYNE FOLIOSA.
Euphrosyne foliosa, *Audouin & M.-Edwards, Ann. Sc. Nat.* tom. xxviii. p. 201, tab. ix. f. 1–15, *Littoral de la France,* ii. 126, tab. ii B. f. 1–14; *M.-Edwards, Cuv. R. An.* ed. *Crochard* (plate), tab. viii. f. 2; *W. Thompson, Ann. & Mag. Nat. Hist.* 2nd ser. tom. iii. p. 355, 1849.
Euphrosine foliosa, *M.-Edwards, Cuv. R. An.* ed. *Croch.* (description of plate); *Quatrefages, Hist. Nat. Annelés,* i. 408 (in synonyms).
Euphrosyne foliosa, *Grube, Famil. der Annelid.* 41 & 122; *Gosse, Ann. & Mag. Nat. Hist.* 2nd ser. tom. xii. p. 384, 1853; *Johnston, Catal. British Non-parasit. Worms,* p. 126; *Carus, Handb. der Zool.* ii. 435; *Ehlers, Die Borstenwürmer,* p. 65; *Quatrefages, Hist. Nat. Annelés,* i. 408.

Hab. Great Britain and Ireland (*Mus. Brit.*); coast of France, *M.-Edwards.*

Sp. 4. EUPHROSYNE POLYBRANCHIA.
Euphrosyne polybranchia, *Schmarda, Neue wirbell. Thiere,* ii. 136, tab. xxxiii. f. 264–287; *Ehlers, Die Borstenwürmer,* p. 65.
Hab. Cape of Good Hope, *Schmarda.*

Sp. 5. EUPHROSYNE CAPENSIS.
Euphrosyne capensis, *Kinberg, Öfvers. af Kongl. Vetensk.-Akad. Förhandl.* 1857, p. 14; *Ehlers, Die Borstenwürmer,* 56.
Hab. Cape of Good Hope, *Kinberg.*

Sp. 6. EUPHROSYNE MEDITERRANEA.
Euphrosyne mediterranea, *Grube, Trosch. Archiv f. Naturgesh.* 1863, tom. xxix. p. 38, tab. iv. f. 2; *Ehlers, Die Borstenwürmer,* p. 66; *Quatrefages, Hist. Nat. Annelés,* i. 409.
?Euphrosyne myrtosa, var. *Ehlers, l. c.*
Euphrosyne mediterranea, *Quatrefages, l. c.* (in synonyms).
Hab. Villa Franca, *Grube.*

Sp. 7. EUPHROSYNE RACEMOSA.
Euphrosyne racemosa, *Ehlers, Die Borstenwürmer,* pp. 66, 67, 80, tab. i. f. 1–11, tab. ii. f. 1, 2.
Hab. Quarnero, Adriatic, *Ehlers.*

Sp. 8. EUPHROSYNE ARMADILLO.
Euphrosyna armadillo, *Sars, Riese i Lofot. og Finmark, Nyt. Magaz. f. Naturvidensk. i Christiania,* 1851, p. 211; *Forhandling. i Vidensk.-Selskab. i Christiania,* 1860, p. 55.

Euphrosyne armadillo, *Ehlers, Die Borstenwürmer,* p. 66.
Hab. Manger, near Bergen, *Sars.*

II. *Some of the branchiæ only ramose.*

Sp. 9. EUPHROSYNE BOREALIS.
Euphrosyna borealis, *Œrsted, in Kröyer Naturhist. Tidssk.* 1842, 113; *Grœnland. Annulat. dorsibranch.* p. 18, 1843, tab. ii. f. 23-27; *Sars, Reise i Lof. Finmark, Nyt Magaz. for Naturvidensk. i Christiania,* 1851, p. 211; *Forhandl. i Vidensk.-Selskab. i Christiania,* 1860, p. 56; *Ray Soc. Reports,* 1845, 324.
Euphrosyne borealis, *Johnston, Cat. Non-parasit. Worms,* p. 127; *Carus, Handb. der Zool.* ii. 435; *Ehlers, Die Borstenwürmer,* p. 66; *Quatrefages, Hist. Nat. Annelés,* i. 408; *Stimpson, Mar. Invert. Grand Manan,* p. 36?; *Grube, Famil. der Annelid.* pp. 41 & 122.
Hab. Britain? (*Mus. Brit.*), Greenland, *Œrsted.*

B. *Segments with cirri only.*

Sp. 10. EUPHROSYNE CIRRATA.
Euphrosyne cirrata, *Sars, Forhandl. i Vidensk.-Selskab. i Christiania,* 1860, p. 56.
Euphrosyne cirrata, *Ehlers, Die Borstenwürmer,* p. 67.
Hab. Manger, near Bergen, *Sars.*

Family II. HIPPONOIDÆ.

Amphinomea (part.), *Grube, Schmarda, Ehlers, Quatrefages.*
Amphinomiens (part.), *Audouin & M.-Edwards.*
Amphinomæ (part.), *M.-Edwards in Lamarck.*
Hipponacea, *Carus, Handb. der Zoologie.*

The animals belonging to this family are distinguished from the Amphinomidæ by being destitute of a caruncle or crest on the buccal or cephalic segment, and by having their feet disposed in only one row. Branchiæ are present on the dorsal segments of the body, and are disposed either in tufts, rows, or groups. Some of the genera (*Hipponoë*) possess tentacles, antennæ, and palpi; others (*Lophonota, Didymobranchus*) are destitute of these appendages. The eyes are four (*Hipponoë, Lophonotus*), placed near each other, small.

Genus I. HIPPONOË.

Hipponoë, *MM. Audouin & M.-Edwards. Ann. Sc. Nat.* 1st series, tom. xx. p. 156, 1830; *Cuvier, M.-Edwards, Guérin, Grube, Van der Hoeven, Carus, Schmarda, Ehlers, Quatrefages.*

Hipponoa, *Audouin & M.-Edwards, Littoral de la France,* ii. p. 128; *M.-Edwards in Lamarck, Costa, Ann. Sc. Nat.* 1841.

Hipponoë, *Audouin & M.-Edwards, Littoral de la France,* ii. p. 117.

The worms belonging to this genus do not appear to have been studied with care. Their eyes are four in number, though M. Quatrefages distinctly asserts that the animals have none. ("Caput oculis destitutum," *l. c.* p. 410.) MM. Audouin and Milne-Edwards do not mention their eyes at all; nor does there seem to be any indication of them in any of the figures of the only species described by them. On the abdominal surface, where, in the Amphinomidæ, the second or ventral series of feet are placed, are a series, on each side of the body, of what Audouin and M.-Edwards called *pores*. No notice is taken by these authors of any appendages connected with them; and M. Quatrefages, in his description of the only species known to him, calls these pores "depressions," and says, they are "destitute of setæ or hooks." ("In abdomine remi inferi loco, depressio quædam setis uncinisque destituta," *l. c.* p. 410.)

These pores or depressions appear to me to be in reality organs of apprehension, for at the bottom of each are a number of short fleshy looking hooked or slightly uncinate spines. The setæ of the feet are all alike, subulate, slender, simple. They possess only one cirrus.

Sp. 1. Hipponoë Gaudichaudi.

Hipponoë Gaudichaudi, *Audouin & M.-Edwards, Ann. Sc. Nat.* 1st series, tom. xx. p. 159, tab. iii. f. 1-5, 1830; *Cuvier, Règne Animal,* 2nd edit. tom. iii. p. 199, 1830; *Guérin, Iconog. R. An.* (texte), *Annélides,* p. 8, tab. iv. bis. f. 3-3A-3D (copied from *Ann. Sc. Nat.*); *M.-Edwards, Règne Anim.* edit. *Crochard,* tab. viii. f. 4, 4A-B, *Littoral de la France,* ii. 239 (note); *Grube, Famil. der Annelid.* p. 41; *Van der Hoeven, Handb. der Zoologie,* i. 231; *Carus, Handb. der Zoologie,* ii. p. 435; *Quatrefages, Hist. Nat. Annelés,* i. 410; *Valenciennes, MS. in Mus. Paris.*

Hipponoa Gaudichaudi, *M.-Edwards, in Lamk. An. s. Vert.* 2nd edit. tom. v. p. 574.

Hab. Port Jackson, *Gaudichaud* fide *Audouin & M.-Edwards*; Madeira, and amongst barnacles on a log of timber near ?Madeira, *Masson (Mus. Brit.)*; concealed in valves of *Lepas fascicularis,* in lat. S. 16° 0', long. W. 5° 0', *Capt. Baker (Mus. Brit.).*

Head small; eyes four, small, placed near to each other. Tentacle larger than antennæ or palpi. Body attenuated at each extremity and composed of about 32 segments. Branchiæ rather small, arbusculiform. Setæ of feet all alike, slender, subulate, simple, rather long. The organs (*pores* of Audouin & M-.Edwards) placed on each of the abdominal segments, on each side, are composed of a depression with a somewhat corrugated raised border round it, and having in the centre of the depression a series (5 to 7) of short fleshy-looking uncinate setæ or spines.

Are these organs organs of apprehension? The specimens of Audouin and M.-Edwards are simply mentioned as having been found at Port Jackson by M. Gaudichaud. The specimens which we possess in the Collection of the British Museum are described as having been found (some) amongst barnacles on a log of ship timber, others as having been concealed in the valves of *Lepas fascicularis* in lat. 16° 0′ S., long. 5° 0′ W., and a third set as from Madeira. By means of these organs I think it probable that they can attach themselves to other bodies partly as parasites. If this be the case, it is curious that in many of the specimens we possess there should be attached to their under surface small animals which are doubtless parasitic to them.

Sp. 2. HIPPONOË CRANCHII, sp. nov. (Plate VI. figs. 7-14.)

In the Museum Collection are two specimens, collected by Mr. Cranch in the Congo expedition, which appear to me to be undescribed.

The worm is only about half an inch long. The body is fusiform in shape, attenuated at either extremity, and composed of about 20 segments. The head is small, but the mouth is very large compared with the size of the animal. The tentacle, antennæ, and palpi are similar to those of the preceding species. Eyes could not be detected. The branchiæ are larger and are composed of more branchlets than those of *Gaudichaudi*. The setæ of the feet are short, simple, and setaceous. The organs (? of apprehension) on the abdominal segments are much larger than in the preceding species, and the fleshy setæ are much longer.

The size of the worm, the comparative size of the mouth, the branchiæ, and the organs of apprehension, and the shortness of the setæ of the feet distinguish this species very well. The organs (? of apprehension) appear to be more produced than in any of the specimens we possess of *Gaudichaudi*, and the setæ are even more hooked at the point, thus confirming my opinion that these organs are retractile, and that, at times at least, they serve the purpose of prehension.

Genus II. LOPHONOTA.

Lophonota, *Costa, Ann. Sc. Nat.* 2nd ser. xvi. 1841; *Grube, Schmarda, Carus, Ehlers, Quatrefages.*

This genus, proposed by Costa for an Annelid found by him in the Bay of Naples, does not seem to have been seen by any other author since his time. It is characterized by having no caruncle, and nothing in the place of tentacle or antennæ. It has four eyes, at least it possesses four small black spots in the place where the eyes ought to be. The feet are disposed in one row only, the setæ are subulate and simple, and the branchiæ are arbuscular, the filiform branches extended in a transverse series across the back. The animal is furnished with a retractile proboscis, which is surrounded with a sort of fringe.

Sp. 1. LOPHONOTA AUDOUINII.

Lophonota Audouinii, *Costa, Ann. Sc. Nat.* 2nd series, tom. xvi. p. 271, 1841, tab. xiii. f. 1, 1A–1D; *Grube, Famil. der Annelid.* p. 41; *Carus, Handb. der Zoologie,* ii. p. 235; *Quatrefages, Hist. Nat. Annelés,* i. p. 411.

Hab. Bay of Naples, *Costa.*

Genus III. DIDYMOBRANCHUS.

Didymobranchus, *Schmarda, Neue wirbell. Thiere,* 1863; *Carus, Handb. der Zool.; Quatrefages.*

This genus has not apparently been seen by any other observer than Schmarda, who has described two species. It is distinguished by the absence of caruncle and antennæ, by having one row of feet, but having both a dorsal and ventral cirrus and the branchiæ pectinated and disposed each in two separate fascicles.

Sp. 1. DIDYMOBRANCHUS CRYPTOCEPHALUS.

Didymobranchus cryptocephalus, *Schmarda, Neue wirbell. Thiere,* ii. p. 138, tab. xxxiii. f. 262; *Quatrefages, Hist. Nat. des Annelés,* i. p. 411.

Hab. Near Valparaiso, *Schmarda.*

Sp. 2. DIDYMOBRANCHUS MICROCEPHALUS.

Didymobranchus microcephalus, *Schmarda, Neue wirbell. Thiere,* ii. 139, tab. xxxiii. f. 263; *Quatrefages, Hist. Nat. Annelés,* i. 411.

Hab. Coast of Chili, *Schmarda.*

Genera belonging to the Amphinomacea, *their true position in which it is difficult to ascertain in consequence of their being too indistinctly described.*

Genus I. SPINTHER.

Spinther, *Johnston,* 1845; *Van der Hoeven, Carus, Grube, Ehlers.*
Cryptonota, *Stimpson,* 1843; *Carus, Quatrefages.*
Oniscosoma, *Sars, Grube, Carus.*

Sp. 1. SPINTHER ONISCOIDES.

Spinther oniscoides, *Johnston, Ann. & Mag. Nat. Hist.* tom. xvi. p. 9, tab. ii. f. 7-14, *Catalogue of British Non-parasitical Worms,* p. 127, tab. xiv. (vi. in text), f. 7-14; *Van der Hoeven, Handb. der Zoologie,* i. 232; *Carus, Handbuch der Zoologie,* ii. 435; *Grube, Beschr. neuer od. wen. bekannt. Annel. in Archiv f. Naturg.* 1860, p. 74.

Cryptonota citrina, *Stimpson, Marine Invert. of Grand Manan (Smithsonian Contributions to Knowledge),* p. 36, tab. ii. f. 27; *Quatrefages, Hist. Nat. Annelés,* i. 412.

Hab. Belfast Bay, *Thompson*; Grand Manan, *Stimpson.*

Sp. 2. SPINTHER MINIACEUS, *Grube, Beschr. n. od. wen. bekannt. Annel. in Archiv für Naturg.* 1860, p. 74, tab. iii. f. 3, 3A-B; *Carus, Hand. der Zool.* ii. 436.

Hab Trieste, *Grube.*

Sp. 3. SPINTHER ARCTICUS.

Oniscosoma arcticum, *Sars, Reise i Lofoten og Finmarken, in Magazin for Naturvidensk.* 1850, p. 210; *Grube, Archiv für Naturg.* 1860, p. 75; *Carus, Handb. der Zoologie,* ii. 436.

Hab. Norway, *Sars.*

Johnston first considered the genus *Spinther* to belong to the Aphroditacea, but afterwards he says it is more allied to *Euphrosyne*. In his 'Catalogue of British Non-parasitical Worms,' he says it is so similar to the *Euphrosyne borealis* in external appearance, that the identity of the two species may be questioned, p. 127. Sars considers his genus *Oniscosoma* might belong to *Euphrosyne*, only that it wants branchiæ; and Carus and Ehlers both place it amongst the *Amphinomidæ*.

Genus II. ARISTENIA.

Aristenia, *Savigny, Syst des Annélides*; *Blainville, Audouin & M.-Edwards, Grube, Schmarda, Ehlers, Quatrefages.*

Sp. 1. ARISTENIA CONSPURCATA.

Aristenia conspurcata, *Savigny, Syst. des Annélid.* p. 64; *Annélides gravés,* tab. ii. f. 3, 41–44; *Blainville, Dict. Sc. Nat. art.* Vers, p. 453, *Atlas,* tab. viii. f. 2–2A (copied from *Savigny*); *Audouin & M.-Edwards, Littoral de la France,* p. 130. tab. iiB. f. 13–14 ; *Grube, Famil. der Annel.* p. 41 ; *Quatrefages, Hist. Nat. Annelés,* i. 412.

Hab. Red Sea, *Savigny.*

Only one species of this genus has as yet been described. Savigny considers the genus as belonging to the Amphinomacea, and in this belief he is followed by Blainville, Carus, Grube, Schmarda and Ehlers ; but Quatrefages only admits it a place in this group provisionally. MM. Audouin and M.-Edwards think it comes nearer to *Trophonia* (*Siphonostomum*). Its general appearance, as represented by Savigny in his plate, with the existence of branchiæ on its dorsal surface, might readily enough bring it amongst the *Amphinomidæ*.

Genus III. ZOTHEA.

Zothea, *Risso, Audouin & M.-Edwards, Schmarda, Grube, Ehlers.*

Sp. 1. ZOTHEA MERIDIONALIS.

Zothea meridionalis, *Risso, Hist. nat. Europ. mérid.* tom. iv. p. 425: *Audouin & M.-Edwards, Littoral de la France,* ii. 130; *Grube, Famil. der Annelid.* p. 41.

Hab. Maritime Alps, ?Nice, *Risso.*

Audouin and Milne-Edwards, in noticing this genus, say they cannot refer it to any portion of the Amphinomacea, as it is described by Risso as possessing horny mandibles. Though vaguely described, it has been nevertheless placed in this group by Schmarda and Ehlers.

Family III. PALMYRIDÆ*.

Palmyracea, *Kinberg, Fregatt. Eugen. Resa, Annulat.* p. 1?, 1855 ; *Carus, Handbuch* ; *Schmarda, Neue wirb. Thiere.*
Palmyridæ, *Baird, Journ. of Linn. Soc.* ix.
Palmyrea, *Quatrefages, Hist. Nat. Annelés.*

* This family, in Kinberg's arrangement, as set forth by him in the Voyage of the Danish Frigate 'Eugenia,' was placed amongst the Aphroditacea, and contained the genus *Palmyra.* Carus followed this arrangement, and placed in the family the genera *Palmyra, Paleonotus,* and *Bhawania.* As far as regarded the name of the family and the genus *Palmyra,* I had adopted Kinberg's arrangement in my contributions to the Aphroditacea in the 9th volume of this

Chrysopetalea, *Ehlers, Die Borstenwürmer,* 1864.

No caruncle. Eyes four or (?) two. Feet uniramose (except in *Bhawania*?). Only one bundle of setæ on each foot. Branchiæ in form of flat setæ (or *paleæ*) disposed in rows, on each side of the back, on each segment.

A. *Body short, with few segments.*

Genus I. CHRYSOPETALUM.

Chrysopetalum, *Ehlers, Die Borstenwürmer,* 1864; *Quatrefages.*

Feet uniramose, furnished with only one tuft of setæ. Head-lobe with four or (?) two eyes, a tentacle, two antennæ, and two palpi. The first segment of body provided with four cirri on each side; the succeeding segments with a cirrus on each side. Body nearly as broad as long. Branchiæ placed on each segment, on each side of body, disposed in a fan-shaped row of flat setæ or paleæ*. Paleæ broad and rather short.

Sp. 1. CHRYSOPETALUM FRAGILE.

Chrysopetalum fragile, *Ehlers, Die Borstenw.* p. 81, tab. ii. f. 3–10; *Quatrefages, Hist. Nat. Annelés,* i. 291.

Hab. Quarnero, *Ehlers.*

Sp. 2. ? CHRYSOPETALUM DEBILE.

Palmyra debilis, *Grube, Beschr. neuer oder wenig bekannt. Annelid. in Wiegmann's Archiv f. Naturg.* 1855, i. p. 90, tab. iii. f. 3–5; *Carus, Handb. der Zool.* ii. p. 434.

Chrysopetalum debile, *Ehlers, Die Borstenwürm.* p. 81; *Quatrefages, Hist. Nat. Annelés,* i. 296.

Hab. Villa Franca, *Grube.*

Genus II. PALEONOTUS.

Paleonotus, *Schmarda, Neue wirbell. Thiere; Carus, Handb. der Zool.; Ehlers, Die Borstenwürmer; Quatrefages.*

Society's Journal. Ehlers, however, has, I think, satisfactorily shown that this family is more nearly connected with the Amphinomacea; and as he has carefully worked out the family and genera which appertain to it, I propose following his arrangement, and bringing now the family Palmyridæ into the group of Amphinomaceæ.

* These branchiæ, composed of flat bristles, or *paleæ*, as they have been called, are considered by Savigny, in the case of *Palmyra,* to be the setæ of the dorsal row of feet. He placed the genus amongst the Aphroditacea, and in this arrangement he has been followed by Audouin and M.-Edwards, Grube, &c., who all take the same view of the case with regard to these appendages.

Head-lobe with a tentacle and two antennæ; palpi wanting. Eyes four. First segment of body provided with two cirri on each side, united at the base. Feet uniramose, with only one tuft of bristles. Body oblong, short. Branchiæ as in *Chrysopetalum*; paleæ short and broad.

Sp. 1. PALEONOTUS CHRYSOLEPIS.

Paleonotus chrysolepis, *Schmarda, Neue wirbell. Thiere*, i. 2. p. 163; *Carus, Handb. der Zool.* ii. p. 434; *Ehlers, Die Borstenwürmer*, p. 80; *Quatrefages, Hist. Nat. Annelés*, i. 297.

Hab. Cape of Good Hope, *Schmarda.*

Genus III. PALMYRA.

Palmyra, *Savigny, Blainville, Cuvier, Lamarck, Audouin & M.-Edwards, Grube, Gervais, Van der Hoeven, Kinberg, Carus, Schmarda, Ehlers, Quatrefages.*

Feet uniramose, each foot with two bundles of bristles. Eyes two [*]. Head-lobe with a tentacle and two antennæ. No palpi. First segment of body furnished with two cirri on each side, united at the base. Body short. Branchiæ as in the two preceding genera. Paleæ narrow and rather long.

Palmyra aurifera, *Savigny, Syst. des Annélides*, p. 17; *Blainville, Dict. Sc. Nat. art.* Vers. p. 463; *Lamarck, An. s. Vert.* 1st edit. v. p. 306, 2nd edit. v. p. 541; *Cuvier, Règne Animal*, tom. iii. p. 206; *Audouin & M.-Edwards, Ann. Sc. Nat.* tom. xxvii. p. 445, tab. x. f. 1, *Littoral de la France*, ii. 111, tab. iiA. f. 1–6; *M.-Edwards, Cuv. Règn. An.* ed. Crochard, *Annélides*, tab. xviii. f. 1, 1A, 1B; *Grube, Famil. der Annel.* p. 38; *Van der Hoeven, Handb. der Zool.* i. 232; *Carus, Hand. der Zool.* ii. 434; *Ehlers, Die Borstenwürmer*, p. 80; *Quatrefages, Hist. Nat. des Annelés*, i. 294.

Hab. Isle of France, *Savigny.*

Sp. 2. PALMYRA ELONGATA.

? Palmyra elongata, *Grube & Œrsted, Annulata Œrstediana*, p. 25; *Quatrefages, Hist. Nat. Annelés*, i. p. 298.

Hab. Santa Cruz, *Œrsted.*

Grube, in his description of this species, distinctly says, "eyes four." In his description of "*P. debilis*," he affirms it

---

[*] Quatrefages, in the definition of this genus, says, "Caput oculis destitutum." Savigny distinctly assigns two eyes as a generic character; and Audouin and M.-Edwards, in their 'Littoral de la France,' distinctly affirm the existence of a similar number, and represent the species *P. aurifera* as possessing two of these organs. M.-Edwards gives a similar representation of them in the figure he gives of this same species in Crochard's edition of Cuvier's 'Règne Animal.'

to have only "two eyes." Yet in his remarks on *P. elongata* (*l. c.*), he says, though it differs from *P. aurata* in the form of the ventral setæ, and in the fan of the paleæ covering the back, it is "numero oculorum *P. debili* similior"! I suspect, as in the case of the genus *Hipponoë* (see remarks under the head of this genus), that the eyes, which are small, have not been properly examined, and that, perhaps, in all the species of this family, the eyes are four in number, as Ehlers has so distinctly represented in his figure of *Chrysopetalum fragile* (*l. c.*), and as Schmarda has also done in the figure he gives of his *Paleonotus chrysolepis* (*l. c*).

B. *Body elongate, with numerous segments.*

Genus IV. BHAWANIA.

Bhawania, *Schmarda, Carus, Ehlers, Quatrefages.*

Feet biramose. Body long, with many segments. Head-lobe with a tentacle, two antennæ, and two palpi; branchiæ apparently indistinct. The paleæ numerous, narrow, in shape of spines, disposed in rows. Setæ of dorsal feet broad, obliquely truncate, all the setæ articulate. Eyes ——?

I insert this genus with a doubt. The figure of the only known species, represented by Schmarda, is very different in form from any others of the family. The feet are biramose, which is different also from the typical genera. Schmarda and Ehlers, however, place this genus in this family without any hesitation; and though Quatrefages throws some doubt on the subject, he remarks, " it represents in this family the Aphroditeans with numerous segments (such as some of the *Polynoës* and *Sigalion*), which we have seen to differ in as great a degree from the Aphrodites and the Hermiones" (*l. c.* p. 298).

Sp. 1. BHAWANIA MYRIALEPIS.

Bhawania myrialepis, *Schmarda, Neue wirbell. Thiere,* i. 2. p. 164; *Carus, Handb. der Zool.* ii. 434; *Ehlers, Die Borstenwürmer,* p. 80; *Quatrefages, Hist. Nat. Annelés,* i. p. 297.

Hab. Island of Ceylon, *Schmarda.*

A short account of two hitherto nondescript Annulose Animals of a larval character.

Amongst the species of Annelids in the British Museum were deposited two specimens (in spirits) of annulose animals, which I was led to believe were marine. One had no habitat attached to it; the other was from the Philippine Islands, collected by the late Mr. Cuming. Their general appearance was peculiar, and I was disposed to place them (as Annelids) in a new family, following the family Hipponoïdæ, and to form for them a new genus (a genus of somewhat degraded Annelids) allied in some respects to the Amphinomacea. Like some of the genera belonging to the family Hipponoïdæ, such as *Hipponoë*, they were destitute of caruncle, and had apparently the feet disposed in a single row, whilst, as in *Lophonota*, there was no appearance of either tentacle or antennæ. The branchiæ seemed to be metamorphosed into stellate groups of short setæ placed in rows on the back, where, in the Amphinomacea, the branchiæ are usually placed. Several naturalists to whom I showed these animals at once proclaimed them to be marine; and the general appearance of at least the species figured in Plate V. is such as to lead to that conclusion. Upon more mature examination, however, their resemblance to the larval form of some insects struck me; and in one of the species (Plate VI.), where the head was somewhat more exposed, the larval structure of the organs of the mouth became manifest. By pressing these organs outwards, Mr. Tuffen West was able to make a sketch of them *in situ* (Plate VI. fig. 4); and their resemblance to those of an insect larva struck him forcibly at the time. A more careful examination of the sketch so made tends to show that these are not marine annulose animals, as I was led at first to suppose, but that in reality they must be referred to the larval state of some unknown insects. Their general resemblance, however, to marine animals, and the belief that the structure of no larvæ like these under consideration has ever been published, determined me to bring them before the notice of the Linnean Society; and as Mr. West has given an exact and very good representation of both species along with a good many details, I thought less apology was required.

It is perhaps objectionable to give a generic name to the larval condition of an insect, but in the meantime, till we know something more of the perfect insect to which they belong, and the true nature of these creatures themselves, I have given to them the provisional name of THETISELLA.

The genus may be characterized thus:—a row of tubercles or feet on both sides in a single row, upon which are implanted a tuft of strong setæ. Two (?) pairs of hooks or feet on the ventral surface near the anterior extremity, on the two first thoracic segments (?) A row of short spines disposed in a stellate-formed group along each side of the dorsal surface at a distance from the tubercles or feet. Dorsal surface rough externally.

In the Collection of the British Museum there are two distinct species, both of which I have figured, and of which I beg to append a description. The names, of course, are only provisional.

Sp. 1. THETISELLA FLAVA. (Plate V. figs. 1–11.)

Body of animal of a yellowish colour, (exclusive of setæ) about 1 inch in length, about half as broad as long, stout or convex on the dorsal surface. It is composed of 12 segments, which are very distinctly seen on the dorsal (fig. 1), but very indistinctly marked on the ventral surface. Mouth placed on the ventral surface, but there are no traces of eyes or antennæ. The dorsal surface is rough and covered with very fine granular-looking bodies, interspersed among which are numerous small calcareous spicula (figs. 9 & 11). The ventral surface (fig. 2) is quite smooth, armed near the anterior extremity with *two* pairs (a pair on each side) of curved hook-like bodies, pointing outwardly, like the feet of larvæ (fig. 7), and having along the centre a series (about 6 or 7) of rather large circles surrounded on the outer edge by a raised rim. Apparently there is no depression in the centre of the ring, and no appearance of hooks or setæ. The feet (?) are disposed only in one row. A bundle of setæ or bristles are implanted on the tubercles, which project straight and are rather short and stout; they are rather numerous, stout, flagelliform, rather long, cylindrical for about one half their length, then suddenly and abruptly terminating in a long, slightly curved, capillary, sharp point (fig. 6). Interspersed amongst these there are several flagelliform setæ in each tuft, shorter than the others, with a swollen portion in the middle of the lower and stouter portion (fig. 8). The organs which, at first view, I considered metamorphosed branchiæ consist of a tuft of short spines placed on the dorsal surface of each segment, on each side, about half way betwixt the centre of the back and the feet, and are disposed in a stellate form. Each tuft consists of about from 5 to 7 flattish setæ terminated by a short curved spine (figs. 4, 5).

Length about 14 lines, including the setæ on the feet; breadth about 7 or 8 lines.

Hab. Unknown (Old Collection, *Mus. Brit.*).

Sp. 2. THETISELLA OLIVACEA. (Plate VI. figs. 1–4.)

Body of animal of an olive-colour, short, nearly as broad as long. Setiferous tubercles of feet long, and terminating in a sharp point. The setæ are implanted at various distances upon the tubercles, and present exactly the same character in form as those in *flava*, with a number of the shorter and stouter swollen setæ interspersed among them (Plate VI. figs. 5 & 6). The setæ altogether are nearly double the length of those of *flava*, and the swollen portion of the shorter setæ are very distinctly visible to the naked eye. The segments of the body are about 11 in number, distinct on the back, but indistinct underneath. The dorsal surface (fig. 1) is very rough, with numerous minute granulations. The ventral surface (fig. 2) is smooth and armed on the anterior portion with *two* pairs of similar hook-pointed organs as in *flava*, while the rings in the middle line are distinctly hollow or depressed in the centre. The organs which I at first considered metamorphosed branchiæ are disposed in the same manner as those of *flava*, the setæ or filaments being placed in a stellate form, but having the points straight instead of being curved. The proboscis in this species is partially extruded, and exhibits a fringe of short fleshy tentacles, about nine in number, the centre one being cruciform at the apex (fig. 4) (*vide* description of organs of mouth in explanation of plate).

Length about ½ an inch; breadth of body about 3 lines, but (including setæ) nearly ½ an inch.

Hab. Philippine Islands, *Cuming* (*Mus. Brit.*).

EXPLANATION OF THE PLATES.

PLATE IV. *Setæ of feet of Amphinomidæ.*

Fig. 1 *a*. Seta of dorsal row of feet of *Amphinome rostrata*. × 50 diameters.
Fig. 1 *b*. Seta of ventral row of ditto. × ditto.
Fig. 2 *a*. Dorsal seta of *Amphinome Jukesi*. × ditto.
Fig. 2 *b*. Ventral seta of ditto. × ditto.
Fig. 3 *a*. Dorsal seta of *Hermodice carunculata*. × ditto.
Fig. 3 *b*. Ventral seta of ditto. × ditto.
Fig. 4 *a*. Dorsal seta of *Eurythoë complanata*. × ditto.
Fig. 4 *b*. Ventral seta of ditto. × ditto.
Fig. 5 *a*. Dorsal seta of *Eurythoë clavata*. × ditto.
Fig. 5 *b*. Ventral seta of ditto. × ditto.
Fig. 6 *a*. Dorsal seta of *Lirione Rayneri*. × ditto.
Fig. 6 *b*. Ventral seta of ditto. × ditto.
Fig. 7 *a*. Dorsal seta of *Chloeia tumida*. × ditto.
Fig. 7 *b*. Dorsal seta of ditto, without teeth or serræ. × ditto.
Fig. 7 *c*. Dorsal seta of ditto, of the whole length. × 20 diam.

Fig. 7 *d*. Ventral seta of ditto, of *Chloeia tumida*. × 50 diam.
Fig. 8 *a*. Dorsal seta of *Chloeia parva*. × ditto.
Fig. 8 *b*. Ventral seta of ditto. × ditto.

PLATE V. *Thetisella flava*.

Fig. 1. Dorsal aspect. Animal enlarged two-thirds.
Fig. 2. Ventral aspect. Ditto.
Fig. 2 *a*. Natural size.
Fig. 3. Lateral aspect. × two-thirds.
Fig. 4. Tuft of (?) branchial setæ along with a portion of the skin on which they are seated. × 25 diam.
Fig. 5. A single (?) branchial seta. × 50 diam.
Fig. 6. One of the setæ of the feet. × ditto.
Fig. 7. One of the hooklets on under thoracic surface. × ditto.
Fig. 8. One of the setæ which are found interspersed amongst the ordinary setæ of the feet. × ditto.
Fig. 9. Portion of the dermal surface of dorsal portion of the body, showing the granular structure of the skin, with small round masses and minute calcareous spicula in the intermediate portion. × 100 diam.
Fig. 10. Portion of the lateral surface of body, × two-thirds, and one of the little knobs on its surface. × 25 diam.
Fig. 11. The calcareous spicula interspersed among the dermal scales shown in fig. 9. × 400 diam.

PLATE VI. *Thetisella olivacea* and *Hipponoë Cranchii*.

Fig. 1. *Thetisella olivacea*, dorsal view. × two-thirds.
Fig. 2. *Thetisella olivacea*, ventral view. × ditto.
Fig. 2 *a*. *Thetisella olivacea*. Natural size.
Fig. 3. *Thetisella olivacea*, lateral view. × two-thirds. Similar knobs are present in this species as in the last (Plate V. fig. 10), but they were too much covered with the bundles of setæ to be shown in the figure.
Fig. 4. Head and organs of the mouth, showing the close correspondence of these organs in this animal with those of several larvæ of insects. Supposing these animals to be larvæ, *a, a*, are the antennæ; *m x, m x*, are the maxillæ, with a large inwardly projecting lobe on each, arising from the antebasal point; this lobe has four tactile appendages exactly like a similar process in the larva of the *Clothes Moth*, and two short setæ: *l b r*, is the labrum; *l b*, the labium; *l t, l t*, labial tentacles, composed of a basal joint and two setæ.
Fig. 5. One of the setæ of feet. × 50 diam.
Fig. 6. One of the setæ interspersed among the others. × ditto.
Fig. 7. *Hipponoë Cranchii*, dorsal aspect. × 3 diam.
Fig. 8. *Hipponoë Cranchii*, ventral aspect. × ditto.
Fig. 8 *a*. *Hipponoë Cranchii*. Natural size.
Fig. 9. Head, as seen from beneath. × 25 diam.
Fig. 10. Head, as seen from above. × ditto.
  a, antennæ; *p*, palpi; *t*, tentacle; *p, s, c*, cirrhi.
Fig. 11. Branchial tuft.
Fig. 12. Prehensile organ on ventral surface. × 25 diam.
Fig. 13. One of the setæ of prehensile organ. × 50 diam.
Fig. 14. One of the setæ of feet. × ditto.

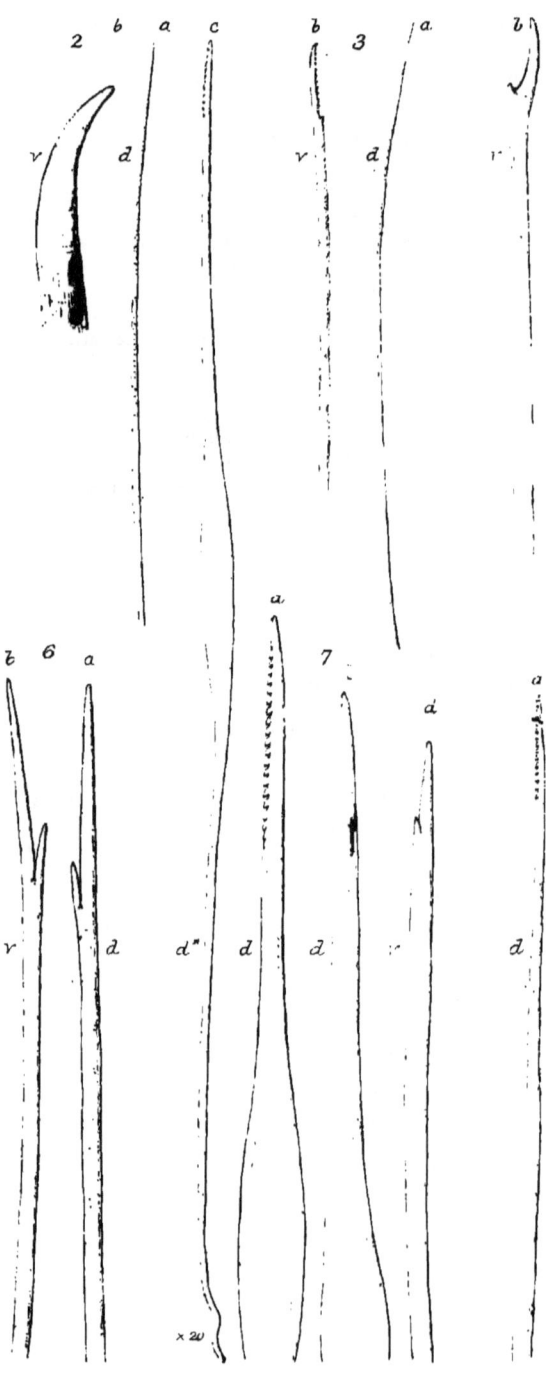

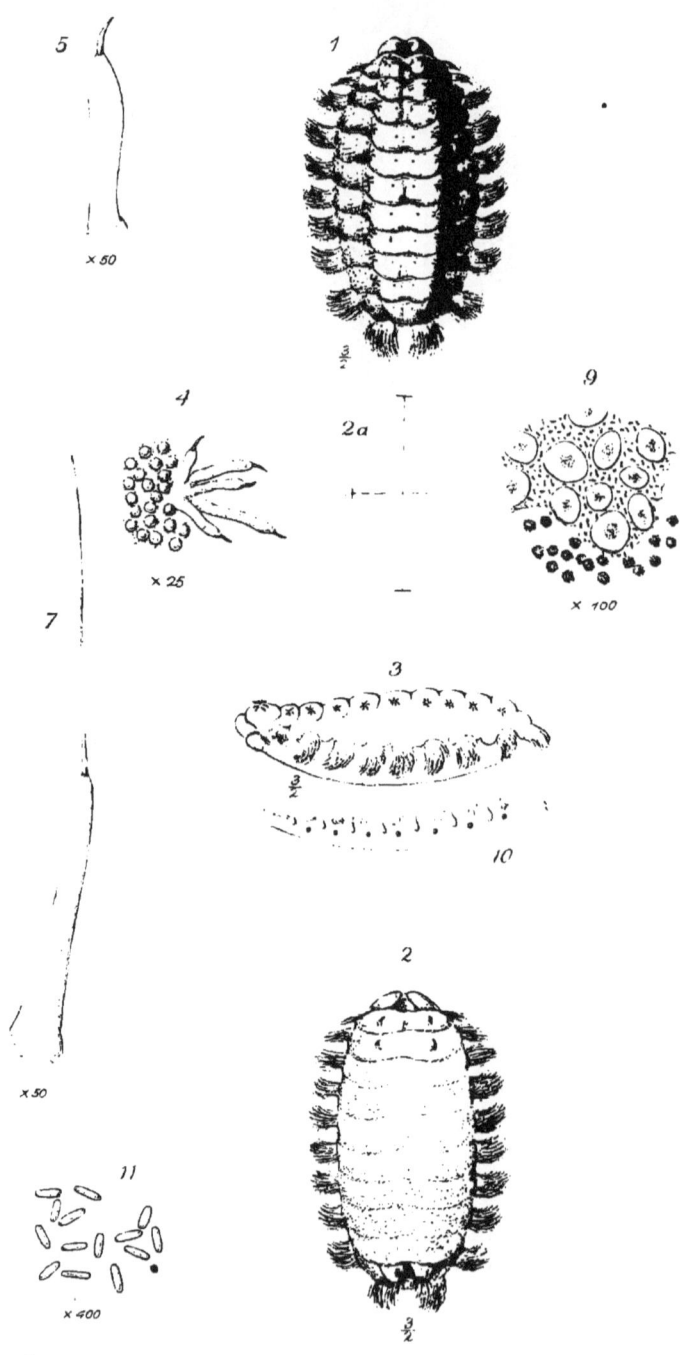

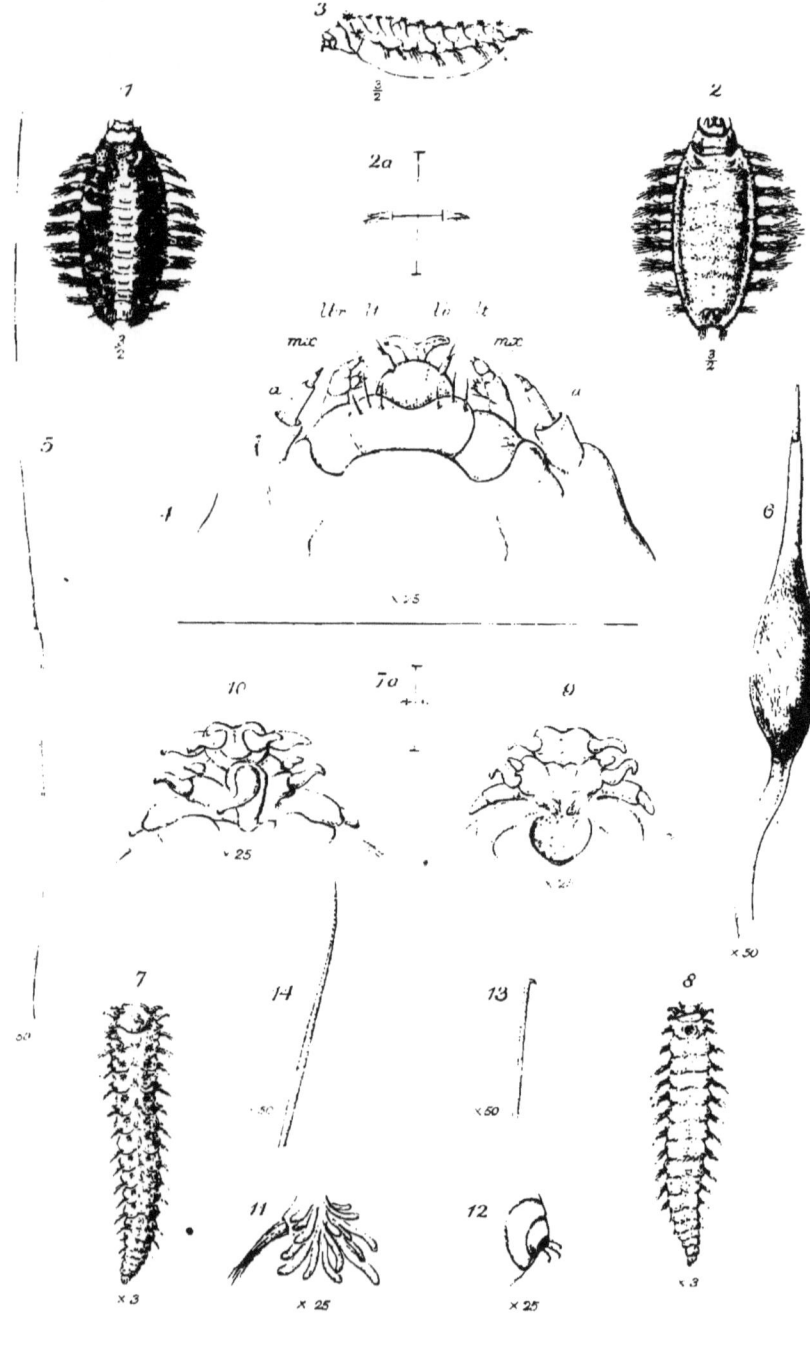

Contributions to a Monograph of the *Aphroditacea*.
By WILLIAM BAIRD, M.D., F.L.S.

(Continued from vol. viii. p. 202.)

[Read Nov. 16, 1865.]

Family IV. ACOETIDÆ.

(Acoëtea, *Kinberg, Öfversigt Kongl. Vetenskaps-Akademiens Förhandlingar*, 1855, p. 386.)

Body elongate; no facial tubercle; tentacle short, arising from the middle of the anterior portion of cephalic lobe; bases of antennæ concealed under the peduncles of the eyes; eyes 2, peduncled; pharynx exsertile, papillose on anterior margin; jaws large, horny, armed with two central and many lateral teeth; palpi long, strong, and smooth. Elytra 39–93 pairs, placed upon the 2nd, 4th, 5th, 7th, and all alternate odd segments onwards to the extremity of the body; segments not bearing elytra provided with dorsal cirri.

Genus I. ACOETES.

Acoëtes, *Audouin & Edwards, Hist. Nat. du Littoral de la France*, ii. p. 92.

Elytra flat, covering the whole back, and arranged imbricately from behind forwards, or in the reverse way to that of the *Aphroditidæ* and *Polynoïdæ*, the posterior portion of each clytron being covered by the anterior of the one behind it; peduncles of eyes about the same length as the peduncle of tentacle.

Sp. 1. ACOETES PLEEI, *Audouin & Edwards, l. c.* p. 101, pl. 2A. figs. 7–14.

Polyodontes Pleei, *Grube, Archiv für Naturg.* 1855, p. 90.

Hab. Martinique, West Indies, *M. Plee*.

Sp. 2. ACOETES LUPINA, *Stimpson, Proceed. Boston Soc.* v. p. 116.

Hab. South Carolina, *Stimpson*.

Genus II. EUPOMPE, *Kinberg, l. c.* p. 386.

Cephalic lobe tripartite on anterior margin; peduncles of eyes a little shorter than peduncle of tentacle, and occupying the anterior portion of cephalic lobe; elytra 93 pairs, flat, thin, and inversely imbricated, or from behind forwards, leaving the anterior and middle portion of the back naked, but covering the posterior part.

Sp. 1. EUPOMPE GRUBEI, *Kinberg, l. c.* p. 387, and in *Fregatten Eugenies Resa*, p. 24, tab. 7. figs. 35A–35H, tab. 10. fig. 59.

Hab. Near Guayaquil, *Kinberg*.

Genus III. PANTHALIS, *Kinberg, l. c.* p. 386.

Cephalic lobe tripartite on anterior margin; central teeth of jaws contiguous; peduncles of eyes of the same length as that of tentacle, and occupying the anterior portion of cephalic lobe. Elytra 39 pairs; the anterior flat, covering the back, inversely imbricated; the remainder campanulate, imbricated posteriorly, and leaving the middle of the back uncovered.

Sp. 1. PANTHALIS OERSTEDI, *Kinberg, l. c.* p. 387, and in *Fregatten Eugenies Resa*, p. 25, tab. 6. figs. 34, 34A–34II, tab. 10. fig. 60.
Hab. West coast of Sweden, *Kinberg.*

Sp. 2. PANTHALIS GRACILIS, *Kinberg, Fregatten Eugenies Resa,* p. 26, tab. 10. fig. 61.
Hab. Near Rio Janeiro, *Kinberg.*

Genus IV. POLYODONTES.

(Renieri) *Blainville,* art. *Vers, Dict. Sc. Nat.* tom. lvii. p. 461.

Elytra very small, not covering the back, alternating with dorsal cirri; jaws large and horny; no antennæ; no tentacle; palpi long; eyes 2, peduncled.

Sp. 1. POLYODONTES MAXILLOSUS.
Phyllodoce maxillosa, *Ranzani, Mem. Stor. Nat. Bologna,* 1820, p. 1, tab. 1. figs. 2–9.
Hab. Adriatic Sea, *Ranzani.*

Sp. 2. POLYODONTES GULO, *Rüppell; Grube, Archiv für Naturg.* 1855, p. 90.
Hab. Red Sea, *Rüppell.*

Family V. SIGALIONIDÆ.

(Sigalionina, *Kinberg, l. e.* p. 387.)

Body long, narrow; no facial tubercle; cephalic lobe rounded. Feet, in anterior segments, provided with either an elytron or a dorsal cirrus; in posterior segments, feet provided with both elytra and dorsal cirri.

Genus I. SIGALION.

Sigalion, *Audouin & Edwards, Hist. Nat. du Litt. de la France,* ii. p. 3 (not *Kinberg*).
Sthenelais, *Kinberg, l. c.* p. 387.

Cephalic lobe rounded, having on its mesial portion, which is indented, a strong tentacle, at the base of which are affixed the antennæ; eyes 2 or 4 (?), sometimes so indistinct as not to be easily seen. Setæ of feet of three kinds—setaceous and serrulate, subulate and serrulate, jointed and bidentate. Elytra covering the back, furnished with simple papillæ.

Kinberg, in taking the *Sigalion Mathildæ* of Audouin and Edwards as the type of his restricted genus *Sigalion*, was, unwittingly perhaps, led into an error, from that species having been originally described by these authors from an imperfect specimen, in which the tentacle was destroyed.

In the illustrations to the 'Règne Animal,' édition Crochard, M. Edwards rectifies this mistake, having, since the first publication of the species, met with other and more perfect specimens: in these the tentacle was present.

It is only right and fair, as Ehlers had already pointed out, to restore the name *Sigalion* to the typical species as correctly described and figured by M. Edwards in the 'Règne Animal.' I therefore propose to retain the name *Sigalion* for the species which Kinberg has placed in his genus *Sthenelais*, and to institute a new genus to receive such as he referred to his restricted genus *Sigalion*.

Sp. 1. SIGALION MATHILDÆ, *Audouin & Edwards, Hist. Nat. du Littoral de la France,* ii. p. 105, tab. 2. figs. 1–10 ; *Règne Animal, éd. Crochard,* tab. 20. figs. 1, 1a–1c.

Hab. Coast of France, *Edwards.*

Sp. 2. SIGALION BOA, *Johnston, Loudon's Mag. Nat. Hist.* vi. p. 322, fig. 42 (1833).

Sigalion Idunæ, *Rathke, Act. Nov. Acad. Nat. Cur.* xx. pt. 1. p. 150, tab. 9. figs. 1–8 (1843).

Hab. Coast of Britain, *Johnston*; Coast of Norway, *Rathke* (*Mus. Brit.*).

Sp. 3. SIGALION HELENÆ.

Sthenelais Helenæ, *Kinberg, l. c.* p. 387, and in *Fregatten Eugenies Resa,* p. 27, tab. 8. figs. 36, 36A–36H.

Hab. Valparaiso, *Kinberg.*

Sp. 4. SIGALION ARTICULATUM.

Sthenelais articulata, *Kinberg, l. c.* p. 387, and in *Fregatten Eugenies Resa,* p. 28, tab. 8. figs. 38, 38A–38H, tab. 10. fig. 62.

Hab. Rio de Janeiro, *Kinberg.*

Sp. 5. SIGALION BLANCHARDI.
Sthenelais Blanchardi, *Kinberg, Fregatten Eugenies Resa*, p. 28, tab. 8. figs. 37A–37H.
Hab. Valparaiso, *Kinberg.*

Sp. 6. SIGALION OCULATUM, *Peters, Monatsbericht Akad. Wissenschaft. Berlin*, 1854, p. 610; *Arch. für Naturg.* 1855, p. 38.
Sthenelais oculata, *Kinberg, Fregatten Eugenies Resa*, p. 29, tab. 8. figs. 39, 39B–39H.
Hab. Mossambique, *Peters.*

Sp. 7. SIGALION LÆVE.
Sthenelais lævis, *Kinberg, Fregatten Eugenies Resa*, p. 29, tab. 8. figs. 40, 40B–40G.
Hab. Island of Eimeo, Pacific, *Kinberg.*

Sp. 8. SIGALION LIMICOLA, *Ehlers, Borstenwürmer*, i. p. 120, tab. 4. figs. 4–7, tab. 5. figs. 1–10.
Hab. Quarnero, Adriatic, *Ehlers.*

Sp. 9. SIGALION ARCTUM?
Aphrodita arcta, *Dalyell, Powers of Creat.* ii. p. 170, tab. 24. fig. 14.
Hab. Coast of Scotland, *Dalyell.*

Sp. 10. ? SIGALION PERGAMENTACEUM, *Grube, Annulata Oerstediana*, p. 24.
Hab. Santa Cruz, West Indies, *Oersted.*
Grube refers this species, with doubt, to the genus *Sigalion.*

Sp. 11. SIGALION BLAINVILLII, *Costa, Ann. Sc. Nat.* 2nd series, xvi. p. 269, tab. 11. figs. 1, 1a–1d.
Hab. Gulf of Naples, *Costa.*

Genus II. THALENESSA.

Sigalion, *Kinberg*, non *Aud. & Edwards.*

Cephalic lobe broad anteriorly; no tentacle; antennæ two, very short, placed on the anterior margin of the cephalic lobe; eyes 2, distant; compound setæ bidentate; simple setæ serrate; elytra covering the back, with ramose fimbriæ on the margin.

Sp. 1. THALENESSA EDWARDSI.
Sigalion Edwardsi, *Kinberg, l. c.* p. 387, and in *Fregatt. Eugen. Resa*, p. 30, tab. 9. figs. 41, 41 A–41 H, t. 10. f. 63.
Hab. Sea off the mouth of the River Plate, South America, *Kinberg.*

Genus III. LEANIRA, *Kinberg, l. c.* p. 388.

Cephalic lobe rounded, receiving the tentacle in a mesial groove; no antennæ; palpi very long; eyes 2, placed near the tentacle; superior setæ closely serrulate; inferior setæ slender, compound, pectinato-canaliculate at the apex; anterior elytra not altogether covering the back; no papillæ.

Sp. 1. LEANIRA QUATREFAGESI, *Kinberg, l. c.* p. 388, and in *Fregatt. Eugen. Resa,* p. 30, tab. 9. figs. 42, 42 A–42 H, tab. 10. fig. 64.

Hab. Sea off the mouth of the River Plate, South America, *Kinberg.*

Sp. 2. LEANIRA STELLIFERA.
Nereis stellifera, *Müller, Zool. Dan.* tab. 62. figs. 1–3.
Sigalion stelliferum, *Sars, Förhand. Vidensk. Selsk. Christiania,* 1861, p. 51.
Sigalion tetragonum, *Oersted, Fortegnelse,* p. 7, tab. 2.
Hab. Coasts of Norway and Sweden, *Müller, Sars, and Oersted.*

Genus IV. PSAMMOLYCE, *Kinberg, l. c.* p. 388.

Cephalic lobe anteriorly produced, and forming the thick base of a long tentacle; antennæ none; eyes 4 ? (2 ?); superior setæ simple, very slender, serrate; inferior setæ strong, bidentate; elytra not covering the middle of the back, with long fimbriæ on their margin.

Sp. 1. PSAMMOLYCE HERMINIÆ.
Sigalion Herminiæ, *Aud. & Edw. Littoral de la France,* ii. p. 107, tab. 1 A. figs. 1–6.
Hab. Rochelle, *M. d'Orbigny.*

Sp. 2. PSAMMOLYCE PETERSI, *Kinberg, l. c.* p. 388, and in *Fregatt. Eugen. Resa,* p. 31, tab. 9. figs. 43, 43 A–43 H.
Hab. Mossambique, *G. v. Düben.*

Sp. 3. PSAMMOLYCE FLAVA, *Kinberg, l. c.* p. 388, and in *Fregatt. Eugen. Resa,* p. 31, tab. 9. figs. 44, 44 A–44 H.
Hab. Rio Janeiro, *Kinberg.*

Genus V. CONCONIA, *Schmarda, Neue wirbell. Thiere,* ii. p. 150.

Segments of body numerous; elytra on 2nd, 4th, 5th, 7th, and all alternate segments up to the 27th, and then on every succeeding segment to the end of the body; dorsal cirri on all the segments. Feet biramous; setæ of upper branch denticulate;

those of inferior branch of two kinds: 1st, simple and strobiliform; 2nd, compound and bidentate. Jaws 4.

Sp. 1. CONCONIA CÆRULEA, *Schmarda, l. c.* tab. 37. fig. 319.
Hab. Coast of Chili, *Schmarda.*

Family VI. PHOLOIDIDÆ.
(Pholoidea, *Kinberg, Fregatt. Eugen. Resa,* p. 1.)

Elytra on all the alternate segments; no dorsal cirri, either on the segments possessing elytra, or on those in which elytra are wanting.

Genus I. PHOLOË, *Johnston, Ann. Nat. Hist.* ii. 428.

Body linear, oblong; proboscis with four horny jaws, the orifice plain; eyes 2; branches of feet connate; bristles of superior branch capillary, those of inferior branch falcate.

Sp. 1. PHOLOË INORNATA, *Johnston, Ann. Nat. Hist.* ii. p. 437, tab. 23. figs. 1–5.
Hab. Cumbrae, Firth of Clyde, *D. Robertson;* Berwick Bay, *Johnston* (*Mus. Brit.*).

Sp. 2. PHOLOË EXIMIA, *Dyster, MS.* in *Johnston's Catalogue of Non-parasitic Worms in British Museum Collection,* p. 122.
Hab. Tenby, *Dyster.*

Sp. 3. PHOLOË BALTICA, *Oersted, Conspect. Annul. Dan.* fascic. i. p. 14, tab. 1. fig. 21, tab. 2. figs. 34–36, 40.
Hab. Coast of Denmark, *Oersted.*

Sp. 4. ? PHOLOË MINUTA, *Oersted, Grœnl. Ann. Dorsib.* p. 17, tab. 1. figs. 3, 4, 8, 9, 16.
Aphrodita minuta, *Fabricius, Faun. Grœnland.* p. 314.
Hab. Godthaab, coast of Greenland, *Oersted.*

Sp. 5. PHOLOË TECTA, *Stimpson, Invertebrata of Grand Manan,* p. 36.
Hab. Grand Manan, in 4 forms, *Stimpson.*

Genus II. GASTROLEPIDIA, *Schmarda, Neue wirbell. Th.* ii. p. 158.

Elytra and dorsal cirri on alternate segments; elytra on 2nd, 4th, 5th, 7th, and all alternate segments up to the 53rd; ventral surface covered on all the segments with elytriform lamellæ; feet biramous.

Sp. 1. GASTROLEPIDIA CLAVIGERA, *Schmarda, l. c.* p. 159, tab. 87. fig. 315.
Hab. Ceylon, *Schmarda.*

Family VII. PALMYRIDÆ.

(Palmyracea, *Kinberg, Fregatt. Eugen. Resa,* p. 1.)

No clytra; fans of flat bristles on all the segments; segments having cirri and tubercles alternately along the back.

Genus I. PALMYRA, *Savigny, Système des Annélides,* p. 16.

Body oblong, depressed; proboscis without tentacles on edge; jaws semicartilaginous; eyes 2; feet with branches separate.

Sp. 1. PALMYRA AURIFERA, *Savigny, l. c.* p. 17.
Hab. Isle of France, *Cuvier*; Red Sea, *Savigny.*

Sp. 2. PALMYRA ELONGATA, *Grube, Annulat. Oersted.* p. 25.
Hab. Santa Cruz, West Indies, *Oersted.*

Sp. 3. PALMYRA DEBILIS, *Grube, Archiv für Naturg.* 1855, p. 90.
Hab. Villa Franca, Mediterranean, *Grube.*

Since this paper on the Aphroditacea was commenced (see vol. viii. of this Journal, p. 172) I have, through the kindness of M. Malmgren, now of Helsingfors, been made acquainted with an excellent paper of his on the Annelides of the North Sea, "Nordiska Hafs-Annulater," published in the 'Öfversight af K. Vet. Akad. Förhandlingar' for 1865. I regret not having seen this paper before these "Contributions to a Monograph of the Aphroditacea" were first commenced in this Journal. In his paper M. Malmgren has instituted no fewer than ten new genera belonging to the family Polynoidæ. Of these I can only here mention the names, with a reference to the species enumerated in my "Contributions."

I. NYCHIA. To this genus he refers nos. 2 & 3 of the genus *Harmothoë,* pp. 194, 195, *H. assimilis* and *H. scabra.* These two species he regards as only one, and as being synonymous with the *Aphrodita cirrosa* of Pallas.

II. EUNOË. To this genus he refers the *Lepidonota scabra* of Oersted, which, upon very good grounds, he considers distinct from the *Aphrodita scabra* of Fabricius.

III. LAGISCA. To this genus he refers no. 11 of the genus *Harmothoë,* p. 195, the *Polynoë rarispina* of Sars.

IV. Evarne. To this genus he refers no. 3 of the genus *Antinoë*, p. 192, the *Polynoë impar* of Johnson.

V. Lanilla. To this genus he refers no. 1 of the genus *Antinoë*, p. 192, the *Polynoë lævis* of MM. Audouin & Edwards.

VI. Melænis, and VII. Eucrantia. Of these two genera no species had been described previously.

VIII. Alentia. To this genus Malmgren refers no 9 of the genus *Halosydna*, p. 187, the *Polynoë gelatinosa* of Sars.

IX. Enipo, and X. Nemidia. These genera approach the restricted genus *Polynoë*; but no species had previously been described.

Some Account of a newly discovered British Fish of the Family *Gadidæ* and the genus *Couchia*. By Jonathan Couch, F.L.S., &c.

[Read Nov. 16, 1865.]

The genus *Couchia* was formed by Mr. W. Thompson, and has been adopted by Dr. Günther, as separated from that of *Motella* or the Rocklings by the more moderately lengthened body of the species, which is also compressed, and by the silvery and brilliant appearance of the sides. In fact, in their general proportions the fishes of this genus are as different from the Rocklings as, among their kindred the other *Gadidæ*, the Pollack and Whiting are from the Ling; while their manners also, so far as they are known, are as different as their shape. And yet, in some of the more prominent particulars of their organization, there exists a similarity between the *Motellæ* and *Couchiæ*, which is the more remarkable as it consists of a relative gradation in the species of each, which is only to be traced throughout by the discovery of one, of which a notice is now presented to the Linnean Society.

As there is a species of *Motella* which is characterized by the presence of four prominent barbs placed in pairs on the front of the head, with a barb dependent also from the lower jaw, so we find in the best-known, and probably most widely spread, of the genus *Couchia*, the Mackerel Midge (*C. glauca*), a similar conformation, together with a characteristic ciliated membrane situated in a chink in advance of the dorsal fin; which membrane certainly is not itself a fin, but an organ of sensibility which is in its most lively motion when the proper fins are at rest. But long before

www.ingramcontent.com/pod-product-compliance
Lightning Source LLC
Chambersburg PA
CBHW020901160426
43192CB00007B/1024